Systems & Control: Foundations & Applications

Series Editor

Tamer Başar, University of Illinois at Urbana-Champaign

Editorial Board

Andrew J. Kurdila
Michael Zabarankin

Convex
Functional Analysis

Birkhauser Verlag
Basel • Boston • Berlin

Authors:

Andrew J. Kurdila
Department of Mechanical and
Aerospace Engineering
University of Florida
Gainesville, FL 32611-6250
USA
ajk@mae.ufl.edu

Michael Zabarankin
Department of Mathematical Sciences
Stevens Institute of Technology
Castle Point on Hudson
Hoboken, NJ 07030
USA
mzabaran@stevens.edu

2000 Mathematics Subject Classification 46N10, 49J15, 49J20, 49J27,
49J40, 49J50, 49K15, 49K20, 49K27, 49N90, 65K10, 90C25, 93C20, 93C25

A CIP catalogue record for this book is available from the Library of
Congress, Washington D.C., USA

Bibliographic information published by Die Deutsche Bibliothek
Die Deutsche Bibliothek lists this publication in the Deutsche
Nationalbibliografie; detailed bibliographic data is available in the Internet
at <http://dnb.ddb.de>.

ISBN 3-7643-2198-9 Birkhäuser Verlag, Basel – Boston – Berlin

© 2005 Birkhäuser Verlag, P.O. Box 133, CH-4010 Basel, Switzerland
Part of Springer Science+Business Media
Printed on acid-free paper produced of chlorine-free pulp. TCF ∞
Printed in Germany
ISBN-10: 3-7643-2198-9
ISBN-13: 978-3-7643-2198-7

9 8 7 6 5 4 3 2 1

Contents

List of Figures

Preface

Overview of Book

This book evolved over a period of years as the authors taught classes in variational calculus and applied functional analysis to graduate students in engineering and mathematics. The book has likewise been influenced by the authors' research programs that have relied on the application of functional analytic principles to problems in variational calculus, mechanics and control theory.

One of the most difficult tasks in preparing to utilize functional, convex, and set-valued analysis in practical problems in engineering and physics is the intimidating number of definitions, lemmas, theorems and propositions that constitute the foundations of functional analysis. It cannot be overemphasized that functional analysis can be a powerful tool for analyzing practical problems in mechanics and physics. However, many academicians and researchers spend their lifetime studying abstract mathematics. It is a demanding field that requires discipline and devotion. It is a trite analogy that mathematics can be viewed as a pyramid of knowledge, that builds layer upon layer as more mathematical structure is put in place. The difficulty lies in the fact that an engineer or scientist typically would like to start somewhere "above the base" of the pyramid. Engineers and scientists are not as concerned, generally speaking, with the subtleties of deriving theorems axiomatically. Rather, they are interested in gaining a working knowledge of the applicability of the theory to their field of interest.

The content and structure of the book reflects the sometimes conflicting requirements of researchers or students who have formal training in either engineering or applied mathematics. Typically, before taking this course, those trained within an engineering discipline might have a working knowledge of fundamental topics in mechanics or control theory. Engineering students may be perfectly comfortable with the notion of the stress distribution in an elastic continuum, or the velocity field in an incompressible flow. The formulation of the equations governing the static equilibrium of elastic bodies, or the structure of the Navier-Stokes Equations for incompressible flow, are often familiar to them. This is usually not the case for first year graduate students trained in applied mathematics. Rather, these students will have some familiarity with real analysis or functional analysis. The fundamental theorems of analysis including the Open Mapping Theorem, the Hahn-Banach Theorem, and the Closed Graph Theorem will constitute the foundations of their training in many cases.

Coupled with this essential disparity in the training to which graduate students in these two disciplines are exposed, it is a fact that formulations and solutions of modern problems in control and mechanics are couched in functional analytic terms. This trend is pervasive.

Thus, the goal of the present text is admittedly ambitious. This text seeks to synthesize topics from abstract analysis with enough recent problems in control theory and mechanics to provide students from both disciplines with a working knowledge of functional analysis.

Organization

This work consists of two volumes. The primary thrust of this series is a discussion of how convex analysis, as a specific subtopic in functional analysis, has served to unify approaches in numerous problems in mechanics and control theory. Every attempt has been made to make the series self-contained.

The first book in this series is dedicated to the fundamentals of convex functional analysis. It presents those aspects of functional analysis that are used in various applications to mechanics and control theory. The purpose of the first volume is essentially two-fold. On one hand, we wish to provide a bare minimum of the theory required to understand the principles of functional, convex and set-valued analysis. We want this presentation to be accessible to those with little advanced graduate mathematics, which makes it a formidable task indeed. For this reason, there are numerous examples and diagrams to provide as intuitive an explanation of the principles as possible. The interested reader is, of course, referred to the numerous excellent texts that present a complete treatment of the theory. On the other hand, we would like to provide a concise summary of definitions and theorems, even for those with a background in graduate mathematics, so that the text is relatively self-contained.

The second book in the series discusses the application of functional analytic principles to contact problems in mechanics, shape optimization problems, control of distributed parameter systems, identification problems in mechanics and control of incompressible flow. While this list of applications is impressive, it is hardly exhaustive. The second volume also overviews recent problems that can be addressed with extensions of convex analysis. These applications include the homogenization of steady state heat conduction equations, approximation theory in identification problems and nonconvex variational problems in mechanics.

The first volume is organized as follows. Chapter 1 begins with a brief overview of topological spaces, and quickly focuses on metric topologies in particular. Two of the most fundamental theorems included in this chapter are the Arzela-Ascoli Theorem and Baire Category Theorem. Next, a brief discussion of normed vector spaces follows. Section 1.4 presents the foundations of measure and integration theory, while Section 1.7 introduces Hilbert Spaces.

Chapter 2 includes a tutorial on bounded linear operator and weak topologies. The chapter presents the Hahn-Banach Theorem, Uniform Boundedness Theorem, Closed Graph Theorem, Riesz's Representation Theorem and common constructions such as Gelfand Triples.

Chapter 3 is particularly important to applications: it includes descriptions of common abstract spaces encountered in practice. A special emphasis is given to a discussion of Sobolev Spaces.

Chapter 4 discusses the most common notions of differentiability for functionals: Gateaux and Fréchet differentiability. It also presents classical examples of differentiable functionals.

Chapter 5 introduces Lagrange multiplier techniques to characterize extrema of optimization problems. The chapter includes a discussion of certain specialized techniques for studying Fréchet differentiable functionals and equality constrained problems.

Chapter 6 represents the fundamentals of classical convex analysis in functional analysis. Convex functionals and sets are defined. The relationship between continuity, convexity and differentiability is outlined. The chapter discusses formal, "engineering techniques" for solving certain optimal control problems. Multiplier methods are derived for convex functionals on ordered vector spaces.

Chapter 7 introduces the important notion of lower semicontinuity and introduces the class of lower semicontinuous functionals. The crucial interplay of semicontinuity and compactness in weak topologies is discussed. The chapter presents several versions of the Generalized Weierstrass Theorem formulated for lower semicontinuous functionals.

Acknowledgements

The authors would like to thank Professor William Hager of the Department of Mathematics, Professor Panos M. Pardalos of the Department of Industrial and Systems Engineering and Professor R. Tyrrel Rockafellar of the Department of Mathematics at the University of Washington for their insight, advice and comments on portions of the manuscript. The authors would also like to thank the numerous project and contract officers who have had the foresight to support various research projects by the authors. Many of the examples discussed in this book are related to or extracted from research carried out under their guidance. Their support has resulted in some of the examples discussed in the text. In particular, the authors would like to acknowledge the generous support of Dr. Kam Ng of the Office of Naval Research, Dr. Walt Silva of NASA Langley Research Center, Dr. Marty Brenner of the Dryden Flight Research Center, and Dr. Gary Anderson of the Army Research Office. Dr. Clifford Rhoades, the director of Physical and Mathematical Sciences of the Air Force Office of Scientific Research, has had a profound influence on the development of this text. Both authors have been fortunate in that they have prepared substantial tracts of the text while working under

AFOSR sponsorship. Furthermore, the authors extends their warmest gratitude to Dr. Pasquale Storza, Director of the University of Florida Graduate Engineering and Research Center (UF-GERC). This book was assembled in its final form while the authors were visiting scholars at UF-GERC.

A large portion of this text was typeset by Maryum Ahmed, Rob Eick, Kristin Fitzpatrick, and Joel Steward, who have our thanks for their careful attention to detail and patience with the authors during the editorial process.

Both authors express their deepest gratitude to their wives and families. The preparation of this book was carried out with their sustained encouragement and support, despite the burden it sometimes put on the families.

Andrew J. Kurdila and Michael Zabarankin
Gainesville, Florida
April 24, 2004

Chapter 1

Classical Abstract Spaces
in Functional Analysis

1.1   Introduction and Notation

Set Theory Notation

Throughout this text, sets are typically denoted by capital letters

$$A,\ B,\ C$$

while elements of sets are denoted by lower case letters

$$a,\ b,\ c.$$

The membership of an element a in a set A is written as

$$a \in A$$

while we write

$$a \notin A$$

if the element a is not in the set A. A set A is sometimes described explicitly by listing its elements

$$A = \{a_1, a_2, \ldots, a_n\}.$$

More frequently, set builder notation is employed wherein we write

$$A = \{a \in X : D(a)\}.$$

That is, A is comprised of all elements a in some well-defined, universal set X such that the description, or property, $D(a)$ is true. For example, if we want to define the set consisting of the positive integers we can write

$$\mathbb{N} = \{i \in \mathbb{R} : i \text{ is an integer and } i > 0\}.$$

As usual, we define the union of sets A and B as all the elements that are a member of A or B:

$$A \cup B = \{c : c \in A \ \text{ or } \ c \in B\}.$$

The intersection of A and B is the collection of elements that are members of both A and B

$$A \cap B = \{c : c \in A \ \text{ and } \ c \in B\}.$$

A set B is a subset of A if every element of B is also an element of A

$$B \subseteq A \quad \Longleftrightarrow \quad (x \in B \Rightarrow x \in A).$$

If B is a subset of A, the complement of B in A, denoted $\tilde{B}$, consists of all the elements of A that are not elements of B. We write

$$\tilde{B} = \{x \in A : x \notin B\}.$$

Sometimes, when we want to emphasize that we refer to the complement of B in the particular set A, we write

$$A \backslash B = \{x \in A : x \notin B\}.$$

Two sets are equal if they are comprised of the same elements. That is, $A = B$ if and only if

$$A \subseteq B \quad \text{and} \quad B \subseteq A.$$

A set B is a proper subset of A if and only if

$$B \subseteq A \quad \text{and} \quad B \neq A.$$

On many occasions, we employ set theoretic operations on families of sets. Suppose $\mathcal{F}$ is a collection of sets. Then we define

$$\bigcup \mathcal{F} \triangleq \{x : x \in F \text{ for some } F \in \mathcal{F}\}$$

$$\bigcap \mathcal{F} \triangleq \{x : x \in F \text{ for all } F \in \mathcal{F}\}.$$

Specific sets occur so frequently in this text that we have reserved designations for them.

$\mathbb{N}$	The set of positive integers
$\mathbb{Z}$	The set of integers
$\mathbb{Q}$	The set of rational numbers
$\mathbb{R}$	The set of real numbers
$\mathbb{C}$	The set of complex numbers

By a sequence in the set X, we mean a mapping either from $\mathbb{N}$ to X or from $\mathbb{Z}$ to X. In this book, sequences will be denoted as

$$\{x_k\}_{k \in \mathbb{N}} \quad \text{or} \quad \{x_m\}_{m \in \mathbb{Z}}$$

depending on the desired index set.

Real Variables

When considering sets consisting of real numbers, conventional notation is employed for the open, closed and half-open intervals

$$[a, b] = \{x \in \mathbb{R} : a \leq x \leq b\}$$
$$(a, b) = \{x \in \mathbb{R} : a < x < b\}$$
$$[a, b) = \{x \in \mathbb{R} : a \leq x < b\}.$$

Of particular importance throughout the text are the infimum, supremum, limit superior and limit inferior of a set of real numbers. If a set of real numbers $A \subseteq \mathbb{R}$ is bounded from above, there is a smallest real number $\bar{a} \in \mathbb{R}$ such that

$$x \leq \bar{a}$$

for all $x \in A$. The number $\bar{a}$ is the supremum of the set $A \subseteq \mathbb{R}$. This relationship is written

$$\bar{a} = \sup\{x : x \in A\}$$
$$= \sup_{x \in A} x.$$

The supremum of a set $A \subseteq \mathbb{R}$ is also sometimes referred to as the least upper bound, or l.u.b., of A. Likewise, if the set A is bounded below, there is a greatest real number $\underline{a} \in \mathbb{R}$ such that

$$\underline{a} \leq x$$

for all $x \in A$. The number $\underline{a}$ is the infimum of A

$$\underline{a} = \inf\{x : x \in A\}$$
$$= \inf_{x \in A} x.$$

The infimum of a set $A \subseteq \mathbb{R}$ is also sometimes referred to as the greatest lower bound, or g.l.b., of A. Let $\{x_n\}_{n \in \mathbb{N}} \subseteq \mathbb{R}$ be some sequence of real numbers. We define the limit superior, or lim sup, of the sequence to be

$$\limsup_{n \to \infty} x_n = \inf_{k \in \mathbb{N}} \left(\sup_{n \geq k} x_n \right).$$

The limit inferior, or lim inf, of the sequence is defined to be

$$\liminf_{n \to \infty} x_n = \sup_{k \in \mathbb{N}} \left(\inf_{n \geq k} x_n \right).$$

The following example illustrates notions of $\limsup_{n \to \infty} x_n$ and $\liminf_{n \to \infty} x_n$.

Example 1.1.1. *Let a sequence $\{x_n\}_{n\in\mathbb{N}}$ be defined as*

$$x_n = \left(1 + \frac{1}{n}\right)\cos \pi n.$$

This sequence is depicted in Figure 1.1. We have

$$
\begin{aligned}
\sup_{n\ge k} x_n &= 1 + \tfrac{1}{k} & \text{for} \quad k = 2j \quad j \in \mathbb{N}\\
\inf_{n\ge k} x_n &= -\left(1 + \tfrac{1}{k}\right) & \text{for} \quad k = 2j - 1 \quad j \in \mathbb{N}.
\end{aligned}
$$

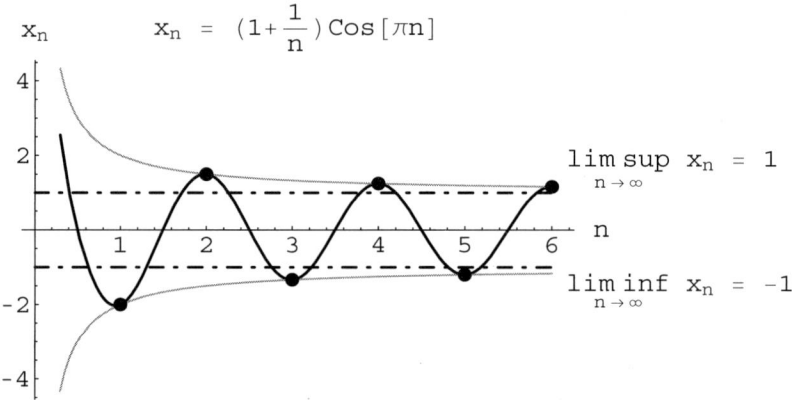

Figure 1.1: Illustrations of notions of $\limsup\limits_{n\to\infty} x_n$ and $\liminf\limits_{n\to\infty} x_n$

By definition of $\limsup\limits_{n\to\infty} x_n$ *and* $\liminf\limits_{n\to\infty} x_n$*, we obtain*

$$
\limsup_{k\to\infty} = \inf_{j\to\infty} \left(1 + \frac{1}{2j}\right) = 1
$$

$$
\liminf_{k\to\infty} = \sup_{j\to\infty} -\left(1 + \frac{1}{2j - 1}\right) = -1.
$$

This example also makes clear that the lim sup, or lim inf, of a sequence of real numbers are simply the largest, or smallest, accumulation points of the sequence, respectively. Note that if a sequence $\{x_n\}_{n\in\mathbb{N}}$ converges, then

$$\limsup_{n\to\infty} x_n = \liminf_{n\to\infty} x_n.$$

For $f : \mathbb{R} \to \mathbb{R}$ the notation $f(x) = O(x)$ means

$$\limsup_{x \to 0} \left| \frac{f(x)}{x} \right| = c$$

for some constant $c \neq 0$. We write $f(x) = o(x)$ whenever

$$\limsup_{x \to 0} \left| \frac{f(x)}{x} \right| = 0.$$

A sequence of real numbers $\{x_k\}_{k \in \mathbb{N}}$ is said to converge to its limit x^* if for any $\epsilon > 0$ there exists n_ϵ such that for any $k > n_\epsilon$, we have $|x_k - x^*| < \epsilon$, or in terms of "$\delta - \epsilon$" notation

$$\forall \epsilon > 0 \ \exists n_\epsilon \in \mathbb{N}: \quad k > n_\epsilon \quad \Longrightarrow \quad |x_k - x^*| < \epsilon.$$

A function $f(x)$ is said to be continuous at point x_0 if for any $\epsilon > 0$ there exists $\delta > 0$ such that for any $x \in (x_0 - \delta, x_0 + \delta)$, we have $|f(x) - f(x_0)| < \epsilon$, i.e.,

$$\forall \epsilon > 0 \ \exists \delta > 0: \quad x \in (x_0 - \delta, x_0 + \delta) \quad \Longrightarrow \quad |f(x) - f(x_0)| < \epsilon.$$

1.2 Topological Spaces

The previous section discussed notions of infimum, supremum, convergence and continuity for sequence of real numbers and functions of real variables. But can we generalize these notions for other abstract mathematical objects? A large part of mathematical analysis has been devoted to giving a precise definition to the notion of "convergence." In conventional language the word "convergence" has many connotations. We say that a hurricane is "converging on the coast," or that "negotiators are converging" in discussions. When asked, most graduate students would have little difficulty in defining a convergent sequence of numbers. Considerable difficulty arises in defining the convergence of more abstract mathematical objects, such as those, for instance, that represent the states of physical systems. For example, when can we say that a sequence of functions converges "as a whole" to another function? To answer to this question we should specify what we mean by "convergence of a function as a whole." This intuitive notion is hardly a rigorous mathematical definition. We should also know whether a considered set of functions is "rich enough" to contain the limit of the sequence. All these matters lead us to determining a structure of abstract spaces under consideration.

For instance, the set of real numbers when plotted may intuitively be perceived as a continuous, infinitely long line. But this graphical description does not determine $\mathbb{R}$ completely. In fact, if we also plot the set of rational numbers $\mathbb{Q}$, then it also "looks" like a continuous, infinitely long line. However, from classical analysis, we know that not every point on the real line belongs to $\mathbb{Q}$. Indeed, $\sqrt{2}$ is such an example, i.e., $\sqrt{2} \notin \mathbb{Q}$. Moreover, if we consider the sequence of rational numbers $\{q_n\}_{n\in\mathbb{N}} \in \mathbb{Q}$ with $q_n = 10^{-n}[10^n \sqrt{2}]$, where $[p]$ denotes the integer part of number p, then obviously $q_n \to q^* = \sqrt{2}$ and the limit $q^* \notin \mathbb{Q}$. The conclusion is that the space $\mathbb{Q}$ is not "rich enough" to contain limits of all convergent sequences of rational numbers. On the other hand, one of the remarkable properties of the space $\mathbb{R}$, taken for granted, is that the limit of any convergent sequence $\{x_n\}_{n\in\mathbb{N}} \in \mathbb{R}$ always belongs to $\mathbb{R}$. This property is known as completeness, which is one of the most important concepts in functional analysis. But abstract spaces may also have many other properties, additionally characterizing their structures and making them different one from another. Notions of completeness, openness, boundedness, connectedness etc. constitute the foundation of the theory of abstract spaces.

In the previous section, we noticed that notions of "convergence" and "continuity" were based on the idea of a *neighborhood*. In the examples of sequence convergence and function continuity, a neighborhood was implicitly defined by the distance between any two points. However, a general definition of a neighborhood does not assume the existence of any measure of distance. Instead, a neighborhood can axiomatically be defined through a structure of open sets, providing a motivation for the introduction of *topological spaces*.

Simply put, a topological space is an abstract space of mathematical objects and a set of axioms that define the structure of open sets. The collection of open sets, in turn, determines the notions of a neighborhood and convergence in this space.

To appreciate how general our definition of convergence of entities in abstract spaces must be to be useful, consider the sequences of images in Figure 1.2. As we examine this sequence of figures, it appears to the eye that the figures are "focusing in" on an image that represents the limit. But what is the "limiting image," and how do we define the closeness of each of the images? The following examples provide some insight on what may constitute a neighborhood and convergence for various mathematical objects in abstract spaces.

Example 1.2.1. *Suppose we picture a hierarchy of finer and finer rectilinear grids* $\{M_j\}_{j\in\mathbb{N}}$ *over the unit square. Each grid* M_j *is comprised of constituent cubes* $\square_{m,n}^j$, *as pictured in Figure* 1.3 (a). *We define*

$$M_j = \{\square_{m,n}^j : \quad m = 0, \ldots, 2^j - 1; \; n = 0, \ldots, 2^j - 1\} \qquad (1.1)$$

where

$$\square_{m,n}^j = \left\{ (x,y) \in [0,1] \times [0,1] : \frac{m}{2^j} < x < \frac{m+1}{2^j}, \; \frac{n}{2^j} < y < \frac{n+1}{2^j} \right\}.$$

One sufficient condition for the convergence of a sequence of points $\{p_k\}_{k\in\mathbb{N}} = \{(x_k, y_k)\}_{k\in\mathbb{N}}$ *that lie in the unit square might be stated as follows:*

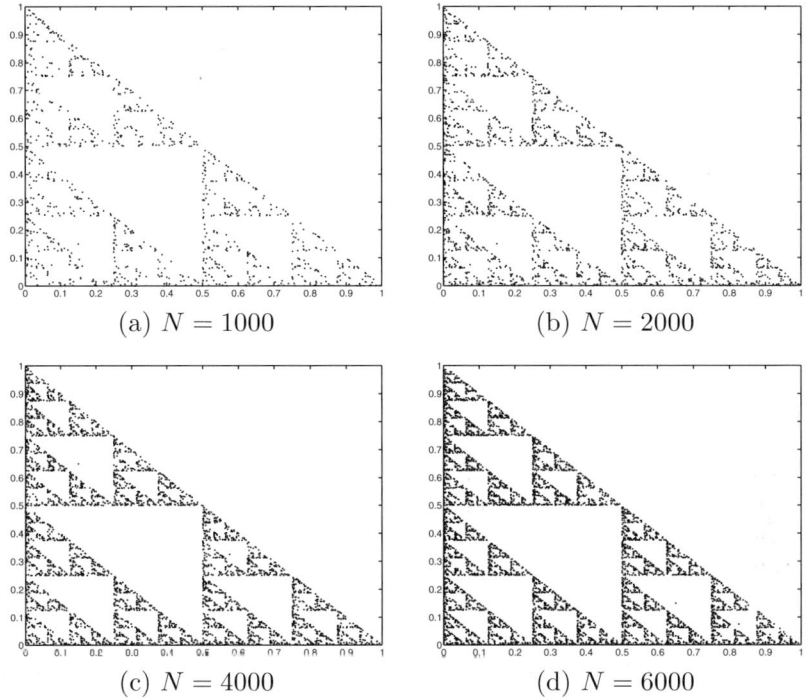

Figure 1.2: Blurred Sierpinski Gasket, Various N

A sequence of points $\{p_k\}_{k\in\mathbb{N}} = \{(x_k, y_k)\}_{k\in\mathbb{N}}$ converges to a point $p^* = (x^*, y^*)$ if for any arbitrarily selected mesh M_j, there is an integer K_j such that the "tail of the sequence" $\{p_k\}_{k\geq K_j}$ lies in the grid box $\square^j_{m^*,n^*}$ that contains p^*.

A moment's reflection will show that this sufficient condition for convergence makes sense independent of the shape of the refined meshes. We could replace the selection of the hierarchy of rectilinear meshes with the triangular meshes as depicted in Figure 1.3 (b). Also note that since this statement embodies a sufficient condition for convergence, there may still be convergent sequences that do not satisfy this condition. Obviously if p^* lies on a dyadic grid line, i.e., $p^* = (i_1 2^{-j_1}, i_2 2^{-j_2})$ for some integers i_1, i_2, j_1 and j_2, then this condition cannot be applicable. Why?

Now, the geometrically intuitive description of convergence given above, and the observation that it should not depend on the "shape" of the grids, leads us to generalize the definition of convergence. If we refer to each of the cubes $\square^j_{m,n}$ comprising grid M_j as a *neighborhood*, the definition can be written more succinctly

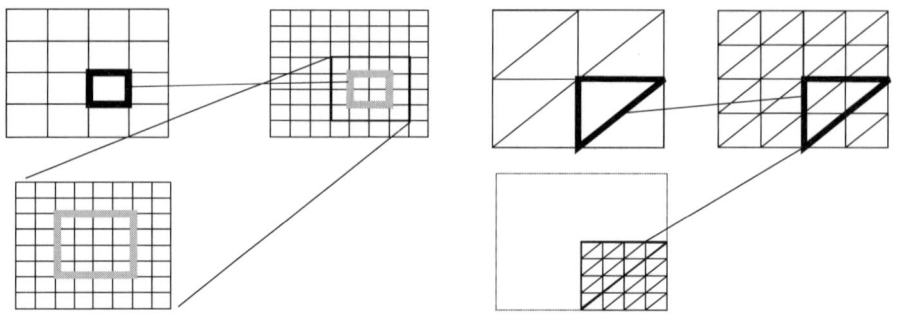

(a) Dyadic Refinements of Unit Square   (b) Triangular Refinements

Figure 1.3: Refinements of a Topological Space

A sequence of points $\{p_k\}_{k\in\mathbb{N}}$ converges to a point p^* if the sequence is *eventually in every neighborhood* of the point p^*.

In this definition, we have used the phrase "eventually in" to replace "the tail of the sequence is contained in." Thus, defining convergence is inextricably connected to the task of defining the points that neighbor a given point.

The following example is fundamental in introductory analysis texts, and discusses the convergence of a classical Fourier series.

Example 1.2.2. *Consider the characteristic function of the unit interval.*

$$f(x) = \mathcal{I}_{\{0\leq x\leq 1\}} = \begin{cases} 1, & x \in [0,1] \\ 0, & x \notin [0,1]. \end{cases}$$

It is well known that we can build approximations of this discontinuous function from very smooth collections of functions. For example, consider the collection

$$\{\phi_k(x)\}_{k\in\mathbb{N}} = \{\sin(k\pi x)\}_{k\in\mathbb{N}}.$$

Suppose we seek a sequence that approximates f on the closed interval $[0,1]$ having the form

$$f_n(x) = \sum_{k=1}^{n} f_k \phi_k(x), \qquad x \in [0,1] \tag{1.2}$$

where the collection of constants f_k, $k \in \mathbb{N}$, must be determined. To find coefficients f_k, we multiply Equation (1.2) by $\phi_m(x)$ and integrate with respect to x to obtain

$$\int_0^1 f_n(x)\phi_m(x)dx = \sum_{k=1}^{n} f_k \int_0^1 \phi_k(x)\phi_m(x)dx. \tag{1.3}$$

Using the relation

$$\sin(k\pi x)\sin(m\pi x) = \frac{1}{2}\left(\cos((k-m)\pi x) - \cos((k+m)\pi x)\right)$$

we obtain

$$\int_0^1 \sin(k\pi x)\sin(m\pi x)dx = \begin{cases} \frac{1}{2} & k = m \\ 0 & k \neq m \end{cases}$$

and consequently, the Equation (1.3) is reduced to

$$\left(-\frac{\cos(m\pi x)}{m\pi} \right)\Bigg|_0^1 = \frac{1}{2}f_m.$$

Finally,

$$f_m = \frac{2}{m\pi}(1 - \cos(m\pi)) = \begin{cases} 0, & m = 2k \\ \frac{4}{\pi m}, & m = 2k + 1 \end{cases}$$

and

$$f_n(x) = \frac{4}{\pi}\sum_{k=0}^{n}\frac{1}{2k+1}\sin((2k+1)\pi x).$$

However, it is well known from classical analysis that the value of the approximants $f_n(x)$ at the ends of the interval $[0, 1]$ do not converge to the values of the function $f(x)$ at 0 or 1. Indeed, for any n, $f_n(0) = 0$ and $f_n(1) = 0$, while $f(0) = f(1) = 1$. On the other hand, the Fourier approximants do appear to approach the function $f(x)$ on $[0, 1]$ as the number of kept terms in the series n gets large, in some sense. In fact, the observation that the approximants do converge to the function "in some sense" led to the definition of convergence "in the mean," that is, L^2 convergence of functions. Consider the simple function as depicted in Figure 1.4(a), and its sequence of Fourier approximants plotted in Figures 1.4(b) through 1.4(d). Clearly, $\{f_n\}_{n\in\mathbb{N}}$ approaches the function $f(x)$ "in the mean." In this case we say that a neighborhood of $f(x)$ on $[0, 1]$ is determined in the L^2 space.

In these examples we noticed that a topological space is defined by stating precisely which subsets of an underlying set constitute the neighborhoods, or open sets, of the topological space. We now make this definition precise.

Definition 1.2.1. *Let X be a set, and let τ be a family of subsets of X. The pair (X, τ) is said to be a topological space if*

(T1) *The empty set is in τ, $\emptyset \in \tau$.*

(T2) *The whole set X is in τ, $X \in \tau$.*

(T3) *If $\{O_\alpha\}_{\alpha \in A}$ is a subfamily (finite or infinite) of sets contained in τ, then $\bigcup_{\alpha \in A} O_\alpha \in \tau$.*

(T4) *If $\{O_\beta\}_{\beta \in A}$ is a finite subfamily of sets contained in τ, then $\bigcap_{\beta \in A} O_\beta \in \tau$.*

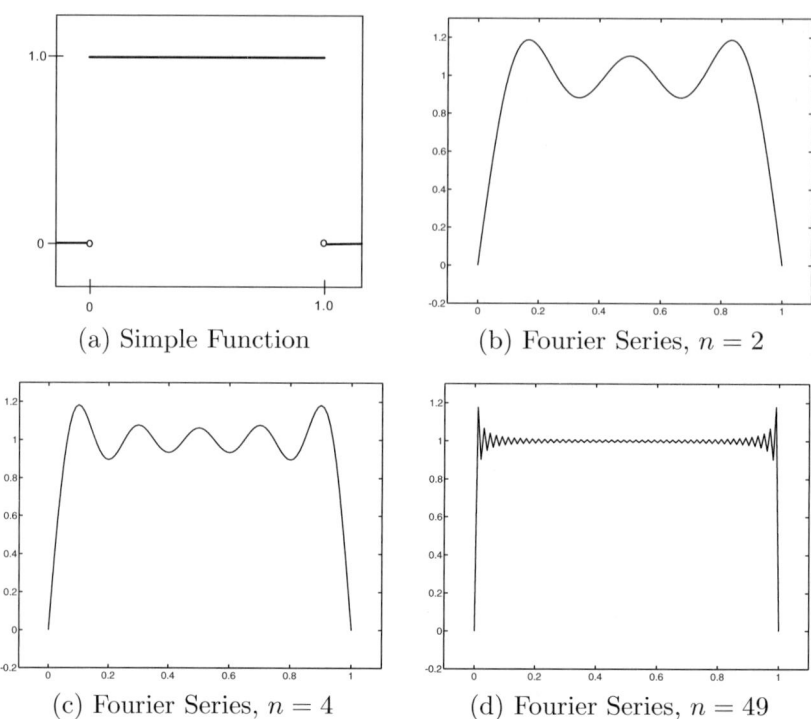

(a) Simple Function (b) Fourier Series, $n = 2$

(c) Fourier Series, $n = 4$ (d) Fourier Series, $n = 49$

Figure 1.4: Simple Function and Sequence of Fourier Approximations

The family of subsets τ of X are the open sets of the topology. A set in the topological space (X, τ) is closed if and only if it is the complement of an open set. That is, C is closed if there exists $O \in \tau$ such that $C = X \backslash O$. Recall that the purpose of introducing a topology on a set is to quantify which points are near other points. A set $\mathcal{N}(x)$ is a neighborhood of the point x if there is an open set O such that $x \in O \subseteq \mathcal{N}(x)$.

Definition 1.2.2. *Let (X, τ) be a topological space and let $A \subseteq X$. The interior of A is the union of all open sets contained in A. The closure of A is the intersection of all closed sets containing A. We denote the interior of A*

$$\text{int}(A) = A^o$$

and the closure of A by

$$\overline{A}.$$

A subset $B \subseteq A$ is said to be dense in A if

$$A \subseteq \overline{B}.$$

Of course, for an arbitrary subset $A \subseteq X$, it need not be the case that $A \in \tau$ or $\overline{A} \in \tau$. In other words, A may not be an open set and A may not be a closed set. In many applications we will be interested in the interior of A and the closure of A.

Definition 1.2.3. *A topological space (X, τ) is separable if and only if it has a countable dense subset.*

Now, in actuality, it is only possible to explicitly list or enumerate all of the open sets that define a topology in the most trivial of examples. One of the easiest ways to define a topology on a set $A \subseteq X$ arises when A inherits its topology from X.

Definition 1.2.4. *Let (X, τ_X) be a topological space. If $A \subseteq X$, the relative topology on A is defined in terms of the open sets*

$$\tau_A = \{B \subseteq A : B = A \cap O \text{ for some } O \in \tau_X\}.$$

The topological space (A, τ_A) is the relativization of (X, τ_X) to the set A.

For topological spaces of interest to engineers and scientists, it is often the case that we can describe sets that are building blocks for all of the open sets that comprise the topology. One means of describing the topology is by describing a *base* for the topology.

Definition 1.2.5. *Let (X, τ) be a topological space, and let $\{B_\alpha\}_{\alpha \in A}$ be a family of subsets of τ. The collection of sets $\{B_\alpha\}_{\alpha \in A}$ is said to form a base for the topological space (X, τ) if any set in τ is the union of sets contained in $\{B_\alpha\}_{\alpha \in A}$.*

The following proposition provides an alternative definition for a base for a topological space.

Proposition 1.2.1. *A collection of sets $\{B_\alpha\}_{\alpha \in A}$ forms a base for a topological space (X, τ) if and only if it satisfies the following two properties*

(1) *any $x \in X$ is contained at least in one B_α,*

(2) *if $x \in B_{\alpha_1} \cap B_{\alpha_2}$ then there exists B_{α_3} such that $x \in B_{\alpha_3} \subset B_{\alpha_1} \cap B_{\alpha_2}$.*

A second useful means of describing the topology on a set gives a description of the open sets that contain a given point. This description is in terms of the *local base* for the topology.

Definition 1.2.6. *Let (X, τ) be a topological space. A family of subsets $\mathcal{B}$ of X is a local base for the topology at the point $x \in X$ if*

(1) *every member of $\mathcal{B}$ is a neighborhood of $x \in X$, and*

(2) *for every neighborhood $\mathcal{N}(x)$ of $x \in X$ there is a set $B \in \mathcal{B}$ such that $B \subseteq \mathcal{N}(x)$.*

Of course, there is a fundamental relationship between a base and a local base for a given topological space.

Theorem 1.2.1. *Let (X, τ) be a topological space. The collection of subsets $\{B_\alpha\}_{\alpha \in A}$ of τ is a base for the topology if and only if $\{B_\alpha\}_{\alpha \in A}$ contains a local base at each $x \in X$.*

Finally, it is often useful to construct a topology from a collection of subsets that cover X, but are not necessarily a base for X.

Definition 1.2.7. *Let X be a set and let $\{O_\alpha\}_{\alpha \in A}$ be a collection of sets that cover X*

$$X \subset \bigcup_{\alpha \in A} O_\alpha.$$

The collection of all finite intersections of sets in $\{O_\alpha\}_{\alpha \in A}$ defines a base for a unique topology on X, referred to as the topology generated by the $\{O_\alpha\}_{\alpha \in A}$.

In general, the index set A for a base $\{B_\alpha\}_{\alpha \in A}$ is not restricted in any way. Of particular interest in applications are the cases in which the index set of a base, or local base, is in fact countable.

Definition 1.2.8. *A topological space (X, τ) is first countable at x if it has a local base at x that is countable. It is first countable if it is first countable at each $x \in X$. The topological space (X, τ) is second countable if it has a countable base.*

Theorem 1.2.1 (Lindelhof's Theorem). *Let the topological space X have a countable base. Every open covering of a set $T \subseteq X$ has a countable subcovering.*

Proof. Let $\{B_i\}_{i \in \mathbb{N}}$ be a countable base for the topology on X and let $\{O_\alpha\}_{\alpha \in A}$ be a covering of T, i.e., $T \subset \bigcup_{\alpha \in A} O_\alpha$. Each $x \in T$ belongs to some O_α. Suppose $x \in O_\alpha$. Since $\{B_j\}_{j \in \mathbb{N}}$ is a base for the topology on X, it contains a local base at each $x \in X$. Consequently, for each $x \in T$ there exists B_j such that $x \in B_j \subset O_\alpha$. If for each B_j we choose $S_j = O_\alpha$, then since base $\{B_i\}_{i \in \mathbb{N}}$ is countable, the subcovering $\{S_i\}_{i \in \mathbb{N}}$ is also countable, and obviously $T \subset \bigcup_{i \in \mathbb{N}} S_i$. $\square$

Example 1.2.3. *Let us pause to re-consider the meshes M_j in equations (1.1). Pick any $x \in [0, 1] \times [0, 1]$ and define a set $B(x)$ such that*

$$B(x) = \{\square_{m,n}^j : x \in \square_{m,n}^j \quad \text{for some } j, m, n \in \mathbb{N}\}.$$

For some choices of x, it is clear that $B(x)$ is not a local base for the usual Euclidean topology on $[0, 1] \times [0, 1]$. If we simply consider any point that lies on one of the mesh lines, say a number like $x = \frac{7}{512}$, $B(x)$ contains a finite number of subsets. On the other hand, if we choose a number $x \in [0, 1] \times [0, 1]$ such that its base-two expansion is non-terminating, $B(x)$ is a local base for the Euclidean topology at x.

1.2.1 Convergence in Topological Spaces

Up to this point, we have emphasized that endowing a set with a topology allows us to describe convergence, and we have noted that convergence can be defined only by introducing a notion of neighborhoods that define "nearness" of elements in a topological space. It turns out that topological spaces often contain rather abstract objects, and in general, it is impossible to index these objects by a sequence of integers. For example, in scientific computation, we often deal with refinements of meshes used in approximating partial differential equations. A prototypical collection of "nested" mesh refinements is depicted in Figures 1.5. Figures 1.5(b) and 1.5(c) depict refinements of the mesh in Figure 1.5(a), but neither is a refinement of the other. Clearly, if we want to number sequentially all such meshes from "the coarsest" to the "finest," such an ordering is not possible. Some meshes are refinements of one another, and are simply incommensurable. Also, we should carefully note that each of the two refinements under consideration do constitute an entry *in a separate path* of refinements that ultimately lead to the most refined mesh depicted in Figure 1.6. Thus, we see that our idea of convergence must also incorporate a more generalized concept of ordering elements in a topological space. This is accomplished by first introducing a *relation* on a set.

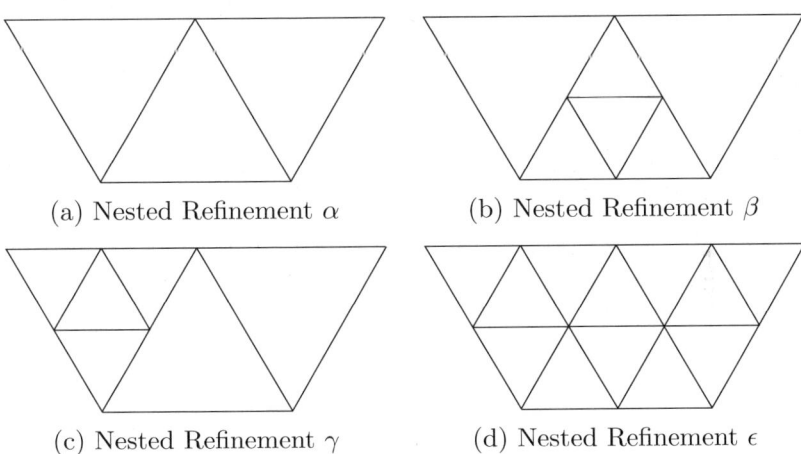

(a) Nested Refinement α (b) Nested Refinement β

(c) Nested Refinement γ (d) Nested Refinement ϵ

Figure 1.5: Nested Refinements

Definition 1.2.9. *Let X be a set. A relation on X is a mapping $R : X \times X \to \{0, 1\}$.*

The interpretation of this definition is actually straightforward; we say that x is *related* to y if $R(x, y) = 1$, otherwise x and y are not related. Another common, frequently employed definition of a relation R defined on a set X simply associates R with a subset of $X \times X$. That is, x and y are related if and only if $(x, y) \in R \subseteq X \times X$. Depending upon the reference, a relation on a set can be denoted in several ways. For example, it is common to employ xRy, $x \leq y$ or $x \geq y$ to indicate that x

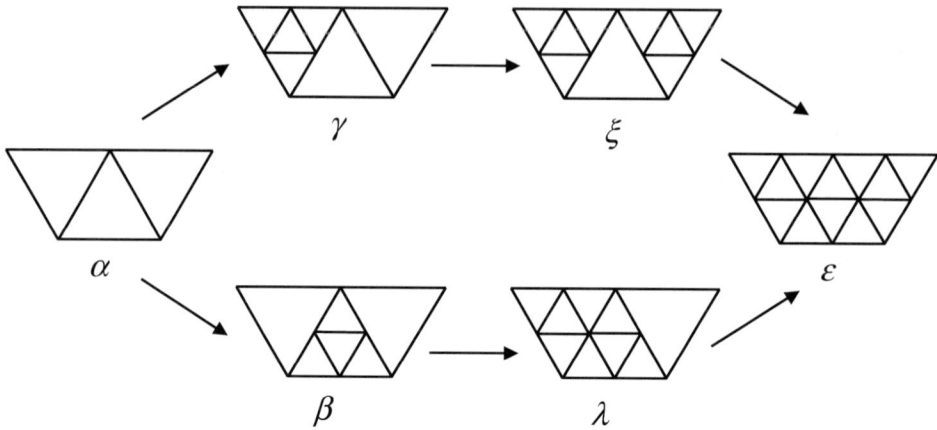

$$\gamma \qquad \xi$$

$$\alpha \qquad\qquad\qquad\qquad\qquad\qquad \varepsilon$$

$$\beta \qquad\qquad \lambda$$

Figure 1.6: Path of Triangular Refinements

is related to y. The important point here is that a relation provides a fundamental construction to order sets. In particular, we will find that a relation serves as the mechanism for defining

- an ordering on a general set X, and
- the equivalence of elements in a general set X.

Thus, depending on the use of the relation, we will use the notation xRy, $x \leq y$, $x \geq y$, or $x = y$ depending on which is most appropriate for the task at hand. We first begin by defining a *partial ordering*, that is, an ordering in which not all elements of the set need be related.

Definition 1.2.10. *Let $\geq$ be a relation on the set X. The set is said to be partially ordered if the relation $\geq$ is*

(PO1)	$x \geq x$	$\forall x \in X$		*Reflexive*
(PO2)	$x \geq y,\ y \geq z$	$\Rightarrow$	$x \geq z$	*Transitive*
(PO3)	$x \geq y,\ y \geq x$	$\Rightarrow$	$x = y$	*Antisymmetric*

A *total ordering* is a partial ordering in which every pair of elements $(x, y) \in X \times X$ are related. Now, the notation in Definition 1.2.10 makes sense. It is trivial to see that the inequality binary operators "$\leq$" and "$\geq$" define total orderings on $\mathbb{R}$. As noted earlier, a relation can also be used to define notions of equivalence on a set X.

Definition 1.2.11. *A relation R defined on a set X induces an equivalence relation on X if R is*

(ER1)	$(x, x) \in R \subseteq X \times X$	$\forall x \in X$		*Reflexive*
(ER2)	$(x, y),\ (y, z) \in R$	$\Rightarrow$	$(x, z) \in R$	*Transitive*
(ER3)	$(x, y) \in R$ if and only if $(y, x) \in R$			*Symmetric*

As a last example of the use of a relation R, we define a directed set. Intuitively speaking, defining a directed set provides a somewhat stronger notion of ordering of a set into subsets that can be ordered by precedence.

Definition 1.2.12. *The pair $(X, \geq)$ consisting of a set X and relation $\geq$ is a directed set if for each x, $y \in X$, there exists a $z \in X$ such that $z \geq x$ and $z \geq y$.*

Finally, we are prepared to define convergence on a general topological space. We define a generalization of sequences that can be used to order abstract elements of a topological space. These generalized sequences are known as *nets*. In some references, they are also called *directions*.

Definition 1.2.13. *A net is a function defined on a directed set. If A is the directed set and the function maps A into X, the net of points in X is denoted $\{x_\alpha\}_{\alpha \in A}$.*

The convergence of a net in a topological space is given by the following definition:

Definition 1.2.14. *A net of points $\{x_\alpha\}_{\alpha \in A}$ is said to converge to a point $x \in X$ if for any neighborhood $\mathcal{N}(x)$ of x, $\exists \alpha_0 \in A$ such that if $\alpha \geq \alpha_0$, then $x_\alpha \in \mathcal{N}(x)$.*

In other words, a net $\{x_\alpha\}_{\alpha \in A}$ converges to some element x if and only if it is eventually in every neighborhood of x. At this point, the reader is urged to compare the abstract Definition 1.2.14 with the heuristic example shown in Figure 1.3. The analogy between the two should be clear. Our previous example is simply a special case of Definition 1.2.14.

1.2.2 Continuity of Functions on Topological Spaces

One of the fundamental tasks is determining what mathematical properties are intrinsic to particular abstract spaces. For topological spaces, concepts like connectedness, continuity and convergence are intrinsic properties that arise from the structure of the space itself. We have spent considerable time discussing *convergence* in topological spaces because it is of fundamental importance in applications. We will also find that *continuity* of functions on topological spaces is likewise of central importance in applications in mechanics and control theory.

Definition 1.2.15. *Let X and Y be topological spaces, a function $f : X \rightarrow Y$ is continuous at point $x \in X$ if for any neighborhood $\mathcal{N}(y) \subset Y$ of point $y = f(x)$, there exists a neighborhood $\mathcal{N}(x) \subset X$ of point x such that $f(\mathcal{N}(x)) \subset \mathcal{N}(y)$.*

This definition may be reformulated in terms of nets, which are widely used in various engineering applications, including optimal control and mechanics. Indeed, relying on our earlier definition of nets, if $\{x_\delta\}_{\delta \in D}$ is a net in X, and if f maps X into Y with $y_\delta = f(x_\delta)$ then $\{y_\delta\}_{\delta \in D}$ is a net in Y. Both nets are indexed by the same directed set D. Continuity of a mapping in a topological space can thus be stated quite simply: every convergent net in the domain is mapped into a convergent net in the range. This definition is easier to remember by recalling one of the definitions of continuity of real-valued functions from calculus:

A function $f : \mathbb{R} \to \mathbb{R}$ is continuous at $x_0 \in \mathbb{R}$ if

$$x \to x_0 \quad \Longrightarrow \quad f(x) \to f(x_0).$$

This result for topological spaces is rigorously proved in the following theorem.

Theorem 1.2.2. *Let X and Y be topological spaces, and let D be a directed set. A function $f : X \to Y$ is continuous at $x \in X$ if and only if*

$$x_\delta \to x \quad \Longrightarrow \quad f(x_\delta) \to f(x).$$

Proof. Sufficiency. If f is a continuous mapping at point x, then for any $\mathcal{Q}_y \subseteq Y$, containing $f(x)$, there exists a neighborhood $\mathcal{Q}_x \subseteq X$ of point x such that $f(\mathcal{Q}_x) \subset \mathcal{Q}_y$. Based on $x_\delta \to x$, there exists δ_0 such that $x_\delta \in \mathcal{Q}_x$ for all $\delta \geq \delta_0$. But for the same $\delta \geq \delta_0$, we have $f(x_\delta) \in \mathcal{Q}_y$, which means that $f(x_\delta) \to f(x)$.

Necessity. Let $x_\delta \to x$ imply $f(x_\delta) \to f(x)$ for every net $\{x_\delta\}_{\delta \in D}$ converging to $x \in X$. Suppose to the contrary that there exists an open set $\mathcal{Q}_y \subseteq Y$ with $f(x) \in \mathcal{Q}_y$ such that for all neighborhoods $\mathcal{Q}_x$ of point x: $f(\mathcal{Q}_x) \not\subset \mathcal{Q}_y$. Let the directed set D be comprised of all neighborhoods of x, and let $\mathcal{Q}_x \geq \mathcal{P}_x$ for any $\mathcal{Q}_x, \mathcal{P}_x \in D$ if $\mathcal{Q}_x \subseteq \mathcal{P}_x$. Define a net on D by choosing $z_{\mathcal{Q}_x} \in \mathcal{Q}_x$ such that $f(z_{\mathcal{Q}_x}) \in f(\mathcal{Q}_x) \backslash \mathcal{Q}_y$. Then $z_{\mathcal{Q}_x} \to x$ but $f(z_{\mathcal{Q}_x}) \nrightarrow f(x)$. $\qquad\square$

There are many alternative characterizations of continuity of functions acting on topological spaces. For the most part, we feel that these characterizations are less intuitive than the definition above. For completeness, we summarize the most common equivalent characterization in the following theorem. It can be very useful in some applications.

Theorem 1.2.3. *Suppose (X, τ_x) and (Y, τ_y) are topological spaces. The mapping $f : X \to Y$ is continuous at $x_0 \in X$ if and only if the inverse image of every open set $\mathcal{Q}_y \subseteq Y$ with $f(x_0) \in \mathcal{Q}_y$ contains an open set $\mathcal{Q}_x \subseteq X$ such that*

$$x_0 \in \mathcal{Q}_x \subseteq f^{-1}(\mathcal{Q}_y).$$

Proof. Necessity. Let f be continuous mapping $X \mapsto Y$ at x_0 and let $\mathcal{Q}_y$ be an open set, containing point $y_0 = f(x_0)$. But it means that $\mathcal{Q}_y$ is a neighborhood of y_0. Consequently, according to Definition 1.2.15, there exists a neighborhood $\mathcal{Q}_x \in X$ of point x_0 such that $f(\mathcal{Q}_x) \subset \mathcal{Q}_y$, that is, $x_0 \in \mathcal{Q}_x \subset f^{-1}(\mathcal{Q}_y)$ and $\mathcal{Q}_x$ is open.

Sufficiency. On the other hand, the fact that for any open set $\mathcal{Q}_y$ with $f(x_0) \in \mathcal{Q}_y$ there is an open set $\mathcal{Q}_x \subset f^{-1}(\mathcal{Q}_y)$ with $x_0 \in \mathcal{Q}_x$ implies that $f(\mathcal{Q}_x) \subset \mathcal{Q}_y$, and consequently f is continuous at x_0. $\qquad\square$

1.2.3   Weak Topology

In applications, one frequently uses continuous functions on topological spaces to define alternative topologies. That is, given two topological spaces X and Y, and a continuous mapping f between them, we can induce a new and sometimes more useful topology on X. This topology is called the *weak topology* from f.

Definition 1.2.16. *Let X be a set, (Y, τ) a topological space, and f a mapping from X into Y, i.e., $f : X \to Y$. The weak topology induced by f on X is defined to be*

$$(X, \text{weak}(f)) \equiv (X, \{O \subseteq X : O = f^{-1}(P) \text{ for some } P \in \tau\}).$$

The practical use of the weak topology, however, usually requires that we induce a topology from some suitably large class of continuous functions defined on X. Obviously, each individual function induces its own weak topology according to Definition 1.2.16. To define a weak topology from a collection of functions, we first consider the more fundamental question of how a topology can be constructed from several auxiliary topologies.

Definition 1.2.17. *Let X be a set and let $\{\tau_\alpha\}_{\alpha \in A}$ be a collection of topologies on X. ((X, τ_α) is a topological space for each $\alpha \in A$). The supremum topology, denoted by*

$$\bigvee_{\alpha \in A} (X, \tau_\alpha)$$

is the topology generated by the collection of subsets $\{\tau_\alpha\}_{\alpha \in A}$ of X.

Definition 1.2.18. *Let X be a set and let $\mathcal{F}$ be a collection of functions such that if $f \in \mathcal{F}$,*

$$f : X \to Y_f.$$

The weak topology on X induced by $\mathcal{F}$ is defined as

$$(X, \text{weak}(\mathcal{F})) \equiv \bigvee_{f \in \mathcal{F}} (X, \text{weak}(f)).$$

Now, this definition has many implications. First of all, the weak topology is the smallest topology for which all of the functions $f \in \mathcal{F}$ are continuous. In addition, it is quite easy to write down precisely which nets converge in X when it is endowed with the weak topology.

Theorem 1.2.4. *Let $\mathcal{F}$ be a collection of functions that induce a weak topology on X, denoted $\big(X, \text{weak}(\mathcal{F})\big)$. A net $\{x_\alpha\}_{\alpha \in A}$ converges in the weak topology on X,*

$$x_\alpha \to x \quad in \quad \big(X, \text{weak}(\mathcal{F})\big)$$

if and only if

$$f(x_\alpha) \to f(x) \quad in \quad Y_f \quad \forall f \in \mathcal{F}.$$

Topological spaces, with no additionally imposed structure, are very abstract indeed. One crucial property of topological spaces X, that is important to engineers and scientists for applications, has to do with the extent to which we can "separate" the points of X. While there are different notions of the degree to which points can be separated in a topological space [34], we need consider only one in this book.

Definition 1.2.19. *A topological space* (X, τ) *is said to be a Hausdorff topological space if for each two distinct points* x, $y \in X$, $x \neq y$, *we can find open subsets* $\mathcal{Q}_x$ *and* $\mathcal{Q}_y$ *such that* $x \in \mathcal{Q}_x$, $y \in \mathcal{Q}_y$ *and*

$$\mathcal{Q}_x \cap \mathcal{Q}_y = \emptyset.$$

The following theorem shows why this property is essential in most applications: it guarantees the uniqueness of limits.

Theorem 1.2.5. *Let* (X, τ) *be a topological space.* X *is a Hausdorff topological space if and only if every convergent net in* X *has a unique limit.*

Proof. Suppose X is a Hausdorff topological space and $\{x_\delta\}_{\delta \in I}$ is a convergent net. Suppose to the contrary that this net converges to both x and y, where $x \neq y$. Since X is a Hausdorff topological space, there are two disjoint open sets $\mathcal{Q}_x$ and $\mathcal{Q}_y$ such that

$$x \in \mathcal{Q}_x \quad \text{and} \quad y \in \mathcal{Q}_y.$$

Since $\{x_\delta\}_{\delta \in I}$ is convergent to x, $\exists \, \delta_x \in I$ such that

$$x_\delta \in \mathcal{Q}_x \qquad \forall \delta \geq \delta_x.$$

Likewise, since $\{x_\delta\}_{\delta \in I}$ is convergent to y, $\exists \, \delta_y \in I$ such that

$$x_\delta \in \mathcal{Q}_y \qquad \forall \delta \geq \delta_y.$$

By the definition of a directed set, there must be some $\delta_0 \in I$ such that

$$\delta_0 \geq \delta_x \quad \text{and} \quad \delta_0 \geq \delta_y.$$

Hence,

$$x_\delta \in \mathcal{Q}_x \cap \mathcal{Q}_y \qquad \forall \delta \geq \delta_0$$

implies that $\mathcal{Q}_x$ and $\mathcal{Q}_y$ are not disjoint, which is a contradiction.

Now suppose that every convergent net in X has a unique limit. Suppose that X is not a Hausdorff topological space. Define a directed set

$$D = \{(\mathcal{Q}_x, \mathcal{Q}_y) : \, \mathcal{Q}_x \text{ is neighborhood of } x, \, \mathcal{Q}_y \text{ is neighborhood of } y\}.$$

We define the relation $\gtrsim$ on D as follows. If

$$(\mathcal{Q}_x, \mathcal{Q}_y) \in D \quad \text{and} \quad (\mathcal{P}_x, \mathcal{P}_y) \in D$$

then

$$(\mathcal{Q}_x, \mathcal{Q}_y) \gtrsim (\mathcal{P}_x, \mathcal{P}_y)$$

if

$$\mathcal{P}_x \supseteq \mathcal{Q}_x \quad \text{and} \quad \mathcal{P}_y \supseteq \mathcal{Q}_y.$$

It is not difficult to show that $(D, \gtrsim)$ defines a directed set. Indeed, if $(\mathcal{Q}_x, \mathcal{Q}_y) \in D$ and $(\mathcal{P}_x, \mathcal{P}_y) \in D$, then

$$(\mathcal{S}_x, \mathcal{S}_y) = (\mathcal{Q}_x \cap \mathcal{P}_x, \mathcal{Q}_y \cap \mathcal{P}_y) \in D$$

and

$$(\mathcal{S}_x, \mathcal{S}_y) \gtrsim (\mathcal{Q}_x, \mathcal{Q}_y)$$
$$(\mathcal{S}_x, \mathcal{S}_y) \gtrsim (\mathcal{P}_x, \mathcal{P}_y).$$

Now, define a net on D by choosing

$$z_{(\mathcal{Q}_x, \mathcal{Q}_y)} \in \mathcal{Q}_x \cap \mathcal{Q}_y$$

for any $(\mathcal{Q}_x, \mathcal{Q}_y) \in D$. The intersection is nonempty by virtue of the fact that X is not a Hausdorff topological space. But the net $\{z_{(\mathcal{Q}_x, \mathcal{Q}_y)}\}_{(\mathcal{Q}_x, \mathcal{Q}_y) \in D}$ converges to both x and y, which is a contradiction. □

1.2.4 Compactness of Sets in Topological Spaces

Finally, we consider what may be the most abstract mathematical property of subsets of a topological space that is required in this volume. To non-mathematicians, the applicability of this concept may not be immediately apparent. However, if the frequency of use is any indication of the importance of a concept, the relevance to application of the definition of a compact set cannot be overstated. Nearly every application in this book series relies on a compactness argument, in one form or another.

Definition 1.2.20. *Let (X, τ) be a topological space. A subset $S \subseteq X$ is compact if and only if every covering of S by open sets contains a finite subcovering.*

Theorem 1.2.6. *Let C be a compact subset in the topological space (X, τ), and let $F \subset C$ be closed, then F is compact.*

Proof. Let $F \subseteq \bigcup_{\alpha \in A} \mathcal{Q}_\alpha$ be a covering of F by open sets. In this case,

$$C \subseteq \left(\bigcup_{\alpha \in A} \mathcal{Q}_\alpha \right) \cup (X \setminus F)$$

defines a covering of C by open sets. Since C is compact, there is a finite collection $\{\mathcal{Q}_{\alpha_k}\}_{k=1}^N$ such that

$$C \subseteq \left(\bigcup_{k=1}^N \mathcal{Q}_{\alpha_k} \right) \cup (X \setminus F).$$

But this implies that

$$F \subseteq \bigcup_{k=1}^{N} \mathcal{Q}_{\alpha_k}$$

and F is consequently compact.   □

The characterization of compactness in terms of a cover of open sets can be intimidating to engineers and scientists. As usual in mathematics, an alternative characterization of compactness is possible. Compactness in topological spaces can be expressed in terms of certain convergent nets. This alternative characterization is made precise by defining cofinal subsets of a directed set and subnets.

Definition 1.2.21. *Let D be a directed set. A subset $E \subseteq D$ is cofinal in D if for each $d \in D$, there is an $e \in E$ such that $e \geq d$.*

It is clear from the definition that a cofinal subset extends throughout a directed set, at least as measured by the order defined on the directed set.

Definition 1.2.22. *Let D, E be directed sets, $\{x_d\}_{d \in D}$ be a net and $f : E \to D$. Then the set*

$$\{x_{f(e)}\}_{e \in E}$$

is a subnet of $\{x_d\}_{d \in D}$ if

1. *$e_1 \geq e_2$ in E implies $f(e_1) > f(e_2)$ in D, and*
2. *$f(E)$ is cofinal in D.*

A little reflection shows that a subnet of a given net is akin to a subsequence of a given sequence. The analogy between subnets and subsequences is fundamental and is made precise in the next section. For now, we simply note that compactness is readily defined in terms of convergent subnets.

Theorem 1.2.7. *A subset C of a topological space is compact if and only if every net contained in C contains a subnet converging to an element in C.*

Thus, we can anticipate an important role for compactness in applications. If a set can be shown to be compact, any extracted net has a convergent subnet. Moreover, it is possible to use the language of nets to show that a set is closed. In applications, one need not produce an open set that is the complement of a given set to show that the given set is a closed set.

Theorem 1.2.8. *A subset C of a topological space is closed if and only if C contains the limit of every convergent net whose elements are contained in C.*

1.3   Metric Spaces

Now that we have defined a topological space, and given some examples for illustration, we will present some of the more common techniques for creating useful topologies. Perhaps the most frequently used method for creating topologies that are of interest to engineers and scientists makes use of *metrics*. In a metric space, we define a distance function that precisely determines what are the open sets, or equivalently the neighborhoods, that comprise the topology. The prototypical metric space is just the real line $\mathbb{R}$ along with the function

$$d(x, y) \equiv |x - y| \quad \forall\, x,\, y \in \mathbb{R}.$$

An arbitrary metric space is defined axiomatically.

Definition 1.3.1. *A metric space is a pair* (X, d) *where* X *is a set and* d *is a distance function defined on the set. The distance function is defined such that*

(M1) $d : X \to \mathbb{R}_0^+$ *(real-valued, finite and nonnegative)*

(M2) $d(x, y) = 0 \iff x = y$

(M3) $d(x, y) = d(y, x)$ *(symmetry)*

(M4) $d(x, z) \leq d(x, y) + d(y, z)$ *(triangle inequality)*

The above definition makes it clear that every metric space is a topological space. The open sets of the topology induced by the metric are just the collection of subsets that can be expressed as the union of open balls $B_r(x)$ where

$$B_r(x) \equiv \{y \in X : d(x, y) < r\}.$$

In other words, the set $\mathcal{B}$, consisting of all open balls $B_r(x)$, is a base for the metric topology. The set $\mathcal{B}(x)$ defined as

$$\mathcal{B}(x) = \{B_r(x) : r > 0\}$$

is a local base at x for the metric topology on X.

1.3.1   Convergence and Continuity in Metric Spaces

With the added structure of a metric, we can specialize the general definition of convergence of a net in a topological space to define convergence of a sequence in a metric space. This definition should start to look quite familiar.

Definition 1.3.2. *A sequence* $\{x_n\}$ *in a metric space* X *is said to converge if* $\exists\, x \in X$ *such that*

$$\lim_{n \to \infty} d(x_n, x) = 0.$$

An intuitive interpretation of this definition is depicted in Figure 1.7. It is clear that this definition is just a special case of convergence of a net in a topological space. When we recall that a neighborhood of a point x contains an open ball $B_r(x) = \{y \in X : d(x, y) < r\}$, it is evident that

$$\lim_{n \to \infty} d(x_n, x) = 0$$

if and only if the sequence of points $\{x_n\}_{n \in \mathbb{N}}$ is eventually in every neighborhood of x.

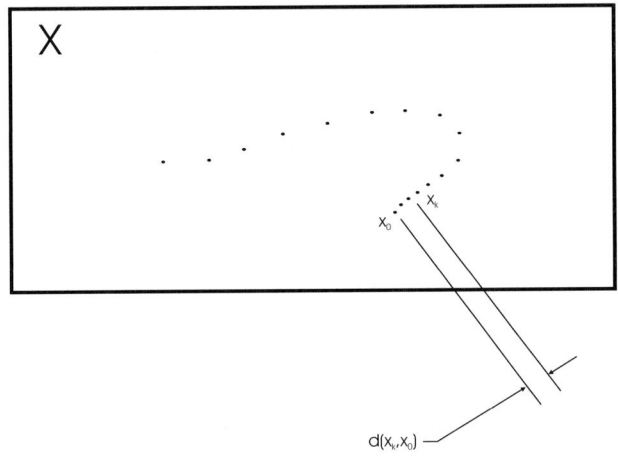

Figure 1.7: Convergent Sequence, Schematic Representation

The definition of continuity of mappings can also be deduced from the definition of continuity of mappings between topological spaces.

Definition 1.3.3. *Let X, Y be metric spaces and consider the mapping $f \colon X \to Y$. The mapping f is continuous at a point $x_0 \in X$ if for every $\epsilon > 0$ there exists $\delta > 0$ such that for any x satisfying $d(x, x_0) < \delta$, we have $d(f(x), f(x_0)) < \epsilon$, or*

$$\forall \epsilon > 0 \; \exists \delta > 0 : \quad d(x, x_0) < \delta \quad \Longrightarrow \quad d(f(x), f(x_0)) < \epsilon.$$

Our statement of this definition is in terms of the familiar "$\delta - \epsilon$" notation, simply because this notation seems to be used most frequently in beginning analysis texts. It should be clear that this definition is equivalent to the statement that

$$x \to x_0 \quad \Longrightarrow \quad f(x) \to f(x_0).$$

That is, the definition essentially says that points in the range of f approach one another as points in the domain do. This observation is consistent with the general definition of continuity in topological spaces.

1.3.2   Closed and Dense Sets in Metric Spaces

We consider next some equivalent notions of closed sets in a metric space. Recall that a set C is closed if and only if it is the complement of an open set in the topology on X. It is seldom the case, however, that this axiomatic definition is particularly useful in applications. A more convenient definition, when we have the added structure of a metric space in place, makes use of the notion of an *accumulation, limit or cluster point*.

Definition 1.3.4. *A point $x \in X$ is a limit (accumulation, cluster) point of $A \subseteq X$ if there is is a sequence of elements $\{x_k\}_{k \in \mathbb{N}} \subseteq A$ such that $x_k \to x$.*

Carefully note that a limit point need not belong to the set A, and that every point in A is trivially a limit point of A (consider the sequence $\{x, x, x, x, \ldots\}$).

Definition 1.3.5. *The closure of a subset A in the metric space X, denoted $\overline{A}$, is the set of all of the limit points of A in X. A set A is closed if and only if $A = \overline{A}$.*

Now that we have available a more useful definition of the closure of a set, we can define when one set "can approximate" another set. The idea that one set is *dense* in another lies at the heart of all problems in approximation theory.

Definition 1.3.6. *The subset A is dense in the metric space X if $\overline{A} = X$.*

When we consider Definitions 1.3.4, 1.3.5, we can fully appreciate the centrality of this definition in approximation. Suppose that a set A is dense in the metric space X, and pick some random element of X. Even if $x \in X$ cannot be "represented explicitly," we are guaranteed that there is a sequence $\{x_k\}_{k \in \mathbb{N}}$ contained in A such that $x_k \to x$. Approximation theory deals with finding sequences $\{x_k\}_{k \in \mathbb{N}}$ defined in terms of "fundamental functions," ones that are easy to compute and are dense in abstract spaces.

1.3.3   Complete Metric Spaces

By adding a metric to a set, we obtain a metric space. It is important to remember that not all topological spaces are metrizable. In addition, another important concept that is introduced with the structure of a metric, that does not arise in a non-metrizable topological space, is the idea of "completeness." Essentially, we can think of a complete metric space as one which does not have any "holes in the limit." This description is not rigorous, of course, but it does convey the spirit of what a complete metric space is. To define a complete metric space, we must first define a Cauchy sequence.

Definition 1.3.7. *A sequence $\{x_n\}_{n \in \mathbb{N}}$ in a metric space X is said to be Cauchy if for every $\epsilon > 0$ there is an $N(\epsilon)$ such that*

$$d(x_m, x_n) < \epsilon, \quad \text{for every} \quad m, n > N(\epsilon).$$

A Cauchy sequence is one for which the elements "squeeze" closer together in the tail of the sequence, as depicted in Figure 1.8. This figure should be compared to Figure 1.7 depicting a convergent sequence. It is important to note that although the elements of the sequence cluster together, *they may not get closer to a single particular point in the set X*. This is what we meant earlier by suggesting that a metric space can have a "hole" in the limit. Thus, we define a complete space as one that has no "holes."

Definition 1.3.8. *The metric space X is said to be complete if every Cauchy sequence in X converges to a point in X.*

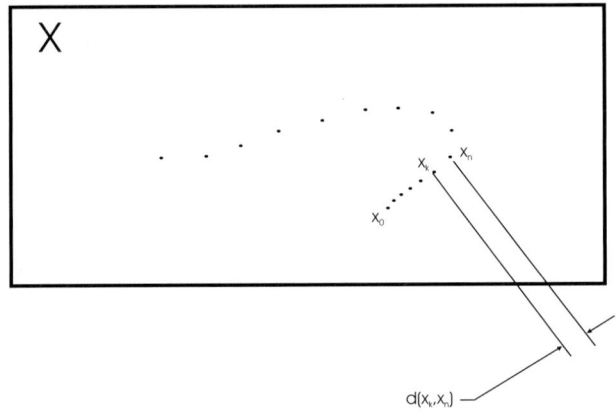

Figure 1.8: Cauchy Sequence, Schematic Representation

Not every metric space is complete. That is, for some metric space X, it is possible to find a sequence $\{x_k\}_{k \in \mathbb{N}}$ that is Cauchy

$$\lim_{k,\,l} d(x_k, x_l) \to 0$$

but does not converge to an element in X.

Example 1.3.1. *The space of all rational numbers $\mathbb{Q}$ is not complete. Indeed, if for any $p, q \in \mathbb{Q}$ we introduce $d(p,q) = |p - q|$ then, clearly, $d(p,q)$ satisfies all the axioms of a metric, but the Cauchy sequence of $q_n = 10^{-n}[10^n \sqrt{2}]$ converges to $\sqrt{2}$, which is not in $\mathbb{Q}$.*

Incomplete metric spaces are, for many applications, flawed in the sense that they do not contain all their points of accumulation. However, it is possible to construct a completion $\hat{X}$ of any metric space X. To define the completion of an arbitrary metric space, we define an isometry i.

Definition 1.3.9. *Let (X, d_X) and (Y, d_Y) be metric spaces. An isometry i is a mapping*

$$i : X \rightarrow Y$$

such that

$$d_X(x_1, x_2) = d_Y(i(x_1), i(x_2)) \qquad \forall\, x_1,\, x_2 \in X.$$

Thus, the isometry "preserves the metric" of the space X. In effect, the metric spaces X_1 and X_2 have the same topological structure. A sequence $\{x_k\}_{k \in \mathbb{N}}$ converges to $x_0 \in X$ in metric if and only if the image of $\{x_k\}_{k \in \mathbb{N}}$ under the isometry $\{i(x_k)\}_{k \in \mathbb{N}}$ converges to $i(x_0) \in Y$. We now can define the completion of a metric space X.

Definition 1.3.10. *Let (X, d_X) be a metric space and let (Y, d_Y) be a complete metric space. The space Y is said to be a completion of X if there exists an isometry i*

$$i : X \rightarrow Y$$

such that the range of $i(X)$ is dense in Y.

The relationship between X, Y and the isometry i is depicted graphically in Figure 1.9. Thus, every metric space can be completed by embedding it isometrically and densely into a complete metric space. While this construction is not difficult, it is lengthy. It is left as an exercise. The full details can be found in numerous texts including [15, 21].

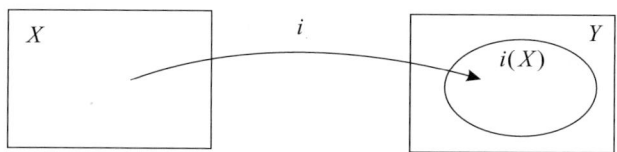

Figure 1.9: Embedding of Incomplete Metric Space X in Complete Metric Space Y via Isometry i

1.3.4 The Baire Category Theorem

The Baire Category Theorem describes the structure of complete metric spaces. Roughly speaking, the Baire Category Theorem asserts that a complete metric space cannot be written as the (countable) union of sets that are too "sparse" or "too thinly populated." Specifically, this notion is made precise in terms of a nowhere dense set.

Definition 1.3.11. *A subset of a topological space X is nowhere dense in X if its closure has empty interior.*

Recall that the interior of a set is the union of all open sets contained in the set. This definition thus implies that if X is a metric space, the closure of a nowhere dense set contains no open ball whatsoever. The Baire Category Theorem then says that we cannot construct a complete metric space in terms of (a countable union) of these meager sets.

Theorem 1.3.1 (The Baire Category Theorem). *A nontrivial complete metric space is not the countable union of nowhere dense sets.*

Proof. Suppose, to the contrary, that X is a complete metric space and

$$X = \bigcup_{i=1}^{\infty} N_i$$

where each N_i is a nowhere dense set. By definition, $\bar{N}_1$ does not contain an open ball. However $X \setminus \bar{N}_1$ is open and nonempty. There is consequently some open ball $B(x_1, r_1) \subseteq X \setminus \bar{N}_1$. Moreover, $X \setminus \bar{N}_2$ is an open set. Since N_2 is nowhere dense, $\bar{N}_2$ does not contain an open set. In particular, $\bar{N}_2$ does not contain the open ball $B(x_1, \frac{1}{2}r_1)$. The intersection $B(x_1, \frac{1}{2}r_1) \cap (X \setminus \bar{N}_2)$ is therefore a nonempty open set. There is some

$$B(x_2, r_2) \subseteq B(x_1, \frac{1}{2}r_1) \cap (X \setminus \bar{N}_2) \subseteq (X \setminus \bar{N}_2).$$

Subsequent iterations yield a nested sequence of open balls

$$B(x_{k+1}, r_{k+1}) \subseteq B(x_k, \frac{1}{2}r_k) \subseteq B(x_k, r_k)$$

$$B(x_k, r_k) \subseteq (X \setminus \bar{N}_k).$$

The sequence $\{x_k\}_{k \in \mathbb{N}}$ is Cauchy, and converges to some $x \in X$ by the completeness of X. In addition, for $n > k$ we have

$$d(x, x_k) \leq d(x_n, x_k) + d(x_n, x).$$

As $n \to \infty$, we obtain

$$d(x, x_k) \leq \frac{1}{2}r_k.$$

Thus,

$$x \in B(x_k, r_k) \qquad \forall k$$

and

$$x \notin N_k \qquad \forall k.$$

Consequently, we conclude that

$$x \notin \bigcup_{k=1}^{\infty} N_k = X.$$

This is a contradiction and the theorem is proved. $\qquad\qquad\square$

1.3.5   Compactness of Sets in Metric Spaces

The next topic in our review of metric spaces is a discussion of compact sets. As we will see later, compactness plays a crucial role in establishing uniform continuity in many applications.

Definition 1.3.12. *A subset A contained in a metric space X is sequentially compact if any sequence extracted from A*

$$\{x_k\}_{k\in\mathbb{N}} \subseteq A$$

contains a subsequence that converges to some $x_0 \in A$. That is, there exists $\{n_k\}_{k\in\mathbb{N}}$ such that

$$x_{n_k} \to x_0 \in A \quad as \quad k \to \infty.$$

A cursory inspection of Definition 1.3.12 would suggest that there might be significant differences between the notion of sequential compactness in a metric space and the general definition of compactness in a topological space in Definition 1.2.20. At best, there is a similarity between sequential compactness in Definition 1.3.12 and the characterization of compactness in terms of convergent subnets in Theorem 1.2.7. However, the following theorem makes clear that all of these definitions coincide on a space whose topology is defined via a metric.

Theorem 1.3.2. *Suppose (X, d) is a metric space. A set $A \subseteq X$ is compact if and only if it is sequentially compact.*

Compact sets play a fundamental role in a host of applications, and are an indispensable tool for engineers and scientists. Indeed, a substantial number of problems in the numerical methods for optimization of partial differential equations are based on showing that a sequence of approximations is contained in a compact set. We can then deduce that at least some subsequence of the approximations converge. In most cases there are still a number of important considerations that must be addressed to solve the problem: for example, is the limit we have obtained actually the solution to the problem? In any event, the use of compactness to find a candidate for the solution is paramount.

As we will see, with a considerable amount of effort, it can be shown that the notion of sequential compactness defined above for metric spaces agrees with the more general definition of compactness in Definition 1.2.20 whenever the topological space is a metric space. We warn the reader, however, that the two definitions of compactness are not equivalent for general topological spaces. Still, the concept of a sequentially compact set plays an important role in many applications of interest to scientists and engineers. To study the relationship of compactness, sequential compactness and boundedness, we introduce two new concepts. We introduce a finite ϵ-net and the definition of a totally bounded set.

Definition 1.3.13. *A finite collection of points $Y_\epsilon \triangleq \{y_1, \ldots, y_n\}$ is said to comprise a finite ϵ-net for a subset A of a metric space (X, d) if for each $x \in A$ there is an* *integer $k \in \{1, \ldots, n\}$ such that*

$$d(x, y_k) \leq \epsilon.$$

Clearly, if Y_ϵ is a finite ϵ-net for the set $A \subseteq X$, the set A is covered by the open balls $B_\epsilon(y_i)$ having radius ϵ and center $y_i \in Y_\epsilon$

$$A \subseteq \bigcup_{i=1}^{n} B_\epsilon(y_i).$$

Finite ϵ-nets are used to define sets that are totally bounded in a metric space.

Definition 1.3.14. *A subset A contained in the metric space (X, d) is said to be totally bounded if for every $\epsilon > 0$ there is a finite ϵ-net for A.*

The concept that a set is totally bounded is stronger than the notion that a set is merely bounded.

Definition 1.3.15. *A precompact set is a totally bounded set in a metric space (X, d).*

Theorem 1.3.1. *Let A be a totally bounded set in a metric space (X, d). Then A is bounded.*

Proof. Pick some $\epsilon > 0$, and let $Y_\epsilon = \{y_1, \ldots, y_n\}$ be an ϵ-net for A. Define

$$\mathcal{C} \triangleq \max_{\substack{i=1, \ldots, n \\ j=1, \ldots, n}} d(y_i, y_j).$$

Choose any two points $x, y \in A$. By the definition of an ϵ-net there are two balls $B_\epsilon(y_m)$ and $B_\epsilon(y_n)$ such that

$$x \in B_\epsilon(y_m)$$
$$y \in B_\epsilon(y_n)$$

where $y_m, y_n \in Y_\epsilon$. By the triangle inequality, we can write

$$d(x, y) \leq d(x, y_m) + d(y_m, y_n) + d(y_n, y)$$
$$\leq \epsilon + \mathcal{C} + \epsilon$$
$$= \mathcal{C} + 2\epsilon.$$

Since $x, y \in A$ are arbitrary, we have that diam $(A) \leq \mathcal{C} + 2\epsilon$ and A is bounded. $\square$

The above result was rather straightforward. Note that the following theorem shows that total boundedness gives a rather concise characterization of sequentially compact sets in metric spaces.

Theorem 1.3.2. *A metric space* (X, d) *is sequentially compact if and only if it is complete and totally bounded.*

Proof. Let $\{x_k\}_{k \in \mathbb{N}}$ be an infinite sequence in X. Suppose that X is complete and totally bounded. Pick $\epsilon = 1$. There is a finite ϵ-net $Y_1 = \{y_{1,1}, \ldots, y_{1,n_1}\}$ for X, and at least one of the open balls $B_1(y_{1,j})$ having radius 1 and center at y_j contains an infinite number of points of the sequence. Choose some $x_{k_1} \in B_1(y_{1,j})$. The set $X \cap B_1(y_{1,j})$ is totally bounded. Pick $\epsilon = \frac{1}{2}$. There is a finite ϵ-net $Y_{\frac{1}{2}} = \{y_{2,1}, \ldots, y_{2,n_2}\}$ for $X \cap B_1(y_{1,j})$, and at least one of the open balls $B_{\frac{1}{2}}(y_{2,k})$ contains an infinite number of points of the infinite sequence. Pick $x_{k_2} \in B_{\frac{1}{2}}(y_{2,k})$. Proceeding in this way, we construct a subsequence $\{x_{k_j}\}_{j \in \mathbb{N}} \subseteq \{x_k\}_{k \in \mathbb{N}}$ and a sequence of nested balls

$$B_1(y_{1,j}) \supset B_{\frac{1}{2}}(y_{2,k}) \supset B_{\frac{1}{4}}(y_{3,l}) \quad \ldots$$

such that

$$x_{k_1} \in B_1(y_{1,j}), \quad x_{k_2} \in B_{\frac{1}{2}}(y_{2,k}), \quad x_{k_3} \in B_{\frac{1}{4}}(y_{3,l}) \quad \ldots .$$

Clearly the subsequence $\{x_{k_j}\}_{j \in \mathbb{N}}$ is a Cauchy sequence. Since the metric space is complete, the sequence converges, which shows that X is sequentially compact. Now suppose that (X, d) is sequentially compact and complete, but to the contrary it is not totally bounded. This means that for some $\epsilon > 0$ there is an infinite sequence $\{x_k\}_{k \in \mathbb{N}}$ such that $d(x_k, x_j) > \epsilon$ for all $k, j \in \mathbb{N}$. However, since the space is sequentially compact there must be a subsequence $\{x_{k_j}\}_{j \in \mathbb{N}}$ that is convergent. But every convergent sequence is Cauchy. There is an M_0 such that $m, n > M_0$ implies $d(x_{k_m}, x_{k_n}) < \epsilon$. But this is a contradiction. $\qquad\square$

From this theorem we see that sequential compactness is equivalent to completeness and total boundedness in a metric space. Further, we will often use the notions of *precompact sets* and *relatively compact sets*.

It often occurs that definitions provided by various authors in abstract analysis use identical terms that vary in a subtle ways from one text to another. One example where subtle variations may arise is in the definition of precompact and relatively compact sets. As noted earlier, a precompact set in a metric space is defined to be a set that is totally bounded. This definition is consistent with that in classical monographs such as [14] as well as with more modern treatments as in [10]. On the other hand, a relatively compact set can be defined in a general topological space as follows.

Definition 1.3.16. *A set* A *in a topological space* X *is relatively compact if the closure of* A *is a compact set in* X.

There is another common definition of a relatively compact set, however. By a relatively compact set some authors also mean a set $A \subseteq X$ such that there is a compact set $K \subseteq X$ such that $A \subseteq K$. The next theorem shows that if the space X is a Hausdorff topological space then these two definitions coincide.

Theorem 1.3.3. *Suppose X is a Hausdorff topological space. Then a set $A \subseteq X$ is a relatively compact if and only if there is a compact set $K \subseteq X$ such that $A \subseteq K$.*

Proof. Suppose A is relatively compact in the Hausdorff topological space X. Then the set $K \triangleq \overline{A}$ is compact by definition. Hence we have constructed a compact set $K \subseteq X$ such that $A \subseteq K$. On the other hand, suppose that there is a compact set $K \subseteq X$ such that $A \subseteq K$. A compact subset of a Hausdorff space is necessarily closed, so we have $K = \overline{K}$. It follows that

$$A \subseteq K \quad \Longrightarrow \quad \overline{A} \subseteq \overline{K} \quad \Longrightarrow \quad \overline{A} \subseteq K.$$

Since in a Hausdorff space a closed subset of a compact set is compact, we conclude that $\overline{A}$ is compact and, consequently A is relatively compact. $\square$

The next theorem characterizes relatively compact sets in metric spaces.

Theorem 1.3.4 (Hausdorff). *A set A is relatively compact in a complete metric space X if A is totally bounded.*

We see that a set A is relatively compact in a metric space X if every sequence in A contains a subsequence that converges in X. The point to which the subsequence converges may not be an element of A, however. The following theorem describes the relationship between precompact and relatively compact sets.

Theorem 1.3.5. *Let X be a metric space. A relatively compact set in X is precompact. If X is complete, a precompact set is relatively compact.*

1.3.6 Equicontinuous Functions on Metric Spaces

In some applications, we will need to establish that a collection of real-valued functions is compact in some suitable topology. A useful strategy for deriving convergent sequences of real-valued functions, when endowed with the uniform metric, can be based on the idea of equicontinuity of a family of functions. We first define equicontinuity of a family $\mathcal{F}$ at a point $x_0 \in X$.

Definition 1.3.17. *Let $\mathcal{F}$ be a family of functions each of which maps the metric space (X, d_X) into the metric space (Y, d_Y). The family $\mathcal{F}$ is said to be equicontinuous at a point $x_0 \in X$ if for each $\epsilon > 0$ there is a $\delta > 0$ such that*

$$\forall f \in \mathcal{F}: \quad d_X(x_0, x) < \delta \quad \Longrightarrow \quad d_Y(f(x_0), f(x)) < \epsilon.$$

If the above definition holds for all $x_0 \in X$, we say that the family $\mathcal{F}$ is equicontinuous on X. The following definition introduces uniform equicontinuity.

Definition 1.3.18. *Let $\mathcal{F}$ be a family of continuous functions each of which maps the metric space (X, d_X) into the metric space (Y, d_Y). The family $\mathcal{F}$ is said to be uniformly equicontinuous if for each $\epsilon > 0$ there is a $\delta > 0$ such that*

$$\forall f \in \mathcal{F}, \quad \forall x_1, x_2 \in X: \quad d_X(x_1, x_2) < \delta \quad \Longrightarrow \quad d_Y(f(x_1), f(x_2)) < \epsilon.$$

We first note that if X is compact, equicontinuity and uniform equicontinuity are equivalent.

Theorem 1.3.6. *Let $\mathcal{F}$ be a family of equicontinuous functions each of which maps the metric space (X, d_X) into the metric space (Y, d_Y). Suppose $A \subset X$ is a compact set, then family $\mathcal{F}$ is uniformly equicontinuous on A.*

Proof. Suppose $\mathcal{F}$ is equicontinuous. Fix $\epsilon > 0$. For each $x \in A$ there is a $\delta_x > 0$ such that

$$d_X(x, y) < \delta_x \quad \Longrightarrow \quad d_Y(f(x), f(y)) < \frac{\epsilon}{2}$$

for all $f \in \mathcal{F}$. The collection of open sets $B_{\frac{\delta_x}{2}}(x)$ of radius $\frac{\delta_x}{2}$ and center x is a covering of A by open sets

$$A \subseteq \bigcup_x B_{\frac{\delta_x}{2}}(x).$$

Since X is compact, there is a finite subcovering

$$A \subseteq \bigcup_{i=1}^{n} B_{\frac{\delta_{x_i}}{2}}(x_i).$$

Set $\delta = \min_i \frac{\delta_{x_i}}{2}$. Suppose y_1, y_2 are any two points in X such that $d_X(y_1, y_2) < \delta$. There is some k such that $y_1 \in B_{\frac{\delta_{x_k}}{2}}(x_k)$. In fact, by the triangle inequality $y_2 \in B_{\delta_{x_k}}(x_k)$:

$$d_X(x_k, y_2) \leq \underbrace{d_X(x_k, y_1)}_{\leq \frac{\delta_{x_k}}{2}} + \underbrace{d_X(y_1, y_2)}_{< \delta \leq \frac{\delta_{x_k}}{2}}.$$

Hence, we have

$$d_Y(f(y_1), f(y_2)) \leq \underbrace{d_Y(f(y_1), f(x_k))}_{\leq \frac{\epsilon}{2}} + \underbrace{d_Y(f(x_k), f(y_2))}_{\leq \frac{\epsilon}{2}} \leq \epsilon$$

for all $f \in d_X$, $y_1, y_2 \in X$, and $\mathcal{F}$ is uniformly equicontinuous. $\qquad\square$

Examples 1.3.2 through 1.3.7 present some of the most common rudimentary metric spaces. A distinguished role is played in applications, however, by the space of continuous real-valued functions defined on a compact metric space X. We denote this set

$$C(X) = \{f : X \to \mathbb{R}, \quad \sup_{x \in X} |f(x)| < \infty\}.$$

We leave it to the reader to verify that $C(X)$ is a metric space with metric

$$d : C(X) \times C(X) \to \mathbb{R}$$
$$d(f, g) \triangleq \sup_{x \in X} |f(x) - g(x)|.$$

Theorem 1.3.7. *Let X be a compact metric space, let $\{f_k\}_{k \in \mathbb{N}} \subseteq C(X)$ be an equicontinuous sequence, and let $Q \subseteq X$ be a dense set. If the sequence of real numbers*

$$\left\{f_k(x)\right\}_{k \in \mathbb{N}}$$

is convergent for every $x \in Q$, then the sequence of functions $\{f_k\}_{k \in \mathbb{N}}$ converges uniformly to an element $f_0 \in C(X)$.

Proof. Fix $\epsilon > 0$. Since the sequence $\{f_k\}_{k \in \mathbb{N}}$ is equicontinuous and defined on the compact set X, it is uniformly equicontinuous. This implies that for any $\epsilon > 0$ and $k \in \mathbb{N}$, there is a $\delta > 0$ such that for all x and y satisfying $d(x, y) < \delta$, we have

$$\left|f_k(x) - f_k(y)\right| < \epsilon. \tag{1.4}$$

Since X is a compact metric space, there is a finite covering of X by open balls

$$X = \bigcup_{k=1}^{n} B_\delta(x_k)$$

having radius δ and centers $\{x_1, \ldots, x_n\}$. The set Q is dense in X and therefore there is at least one element $\xi_k \in Q$ such that

$$\xi_k \in B_\delta(x_k) \qquad k = 1, \ldots, n.$$

By hypothesis, each of the n sequences, $1 \leq k \leq n$,

$$\left\{f_m(\xi_k)\right\}_{m \in \mathbb{N}}$$

is a convergent sequence of real numbers. Let x be an arbitrary point in X. There is some ball $B_\delta(x_k)$ such that $x \in B_\delta(x_k)$. Consider two functions f_m and f_n. We have

$$\left|f_m(x) - f_n(x)\right| \leq \underbrace{\left|f_m(x) - f_m(x_k)\right|}_{\text{term 1}} + \underbrace{\left|f_m(x_k) - f_m(\xi_k)\right|}_{\text{term 2}}$$

$$+ \underbrace{\left|f_m(\xi_k) - f_n(\xi_k)\right|}_{\text{term 5}}$$

$$+ \underbrace{\left|f_n(\xi_k) - f_n(x_k)\right|}_{\text{term 3}} + \underbrace{\left|f_n(x_k) - f_n(x)\right|}_{\text{term 4}}$$

$$\leq 5\epsilon.$$

By construction, x, x_k and ξ_k are elements of $B_\delta(x_k)$, and by the equicontinuity condition (1.4) yield terms 1, 2, 3, and 4 that are bounded by ϵ. By hypothesis, the sequence

$$\left\{f_m(\xi_k)\right\}_{m \in \mathbb{N}}$$

is a convergent Cauchy sequence of real numbers. For m, n large enough, term 5 is bounded by ϵ. In all, we have

$$\sup_{x \in X} |f_m(x) - f_n(x)| \leq 5\epsilon$$

for m, n large enough. Hence the sequence $\{f_k\}_{k \in \mathbb{N}}$ is Cauchy, and since $C(X)$ is complete, it converges to some $f_0 \in C(X)$. $\qquad\square$

1.3.7 The Arzela-Ascoli Theorem

Compactness in general topological spaces is discussed in terms of coverings of sets by open sets in the topology. With the introduction of the metric space structure, considerable simplification is achieved. We are able to connect compactness with the behavior of sequences. We characterize compact sets in terms of more intuitive notions like total boundedness. All of these ingredients allow us to state a powerful theorem that characterizes compactness of sets in the space of continuous functions.

Theorem 1.3.8 (The Arzela-Ascoli Theorem). *Let X be a compact metric space. A subset $\mathcal{F} \subseteq C(X)$ of functions is relatively compact in $C(X)$ if and only if $\mathcal{F}$ is bounded and equicontinuous.*

Proof. Suppose $\mathcal{F}$ is relatively compact in $C(X)$. We want to show that $\mathcal{F}$ is bounded and equicontinuous. Suppose to the contrary that $\mathcal{F}$ is not bounded. Then there is a sequence $\{g_k\}_{k \in \mathbb{N}} \subseteq \mathcal{F}$ such that

$$d(g_k, 0) > k$$

for $k \in \mathbb{N}$. In this case, there is no subsequence of $\{g_k\}_{k \in \mathbb{N}}$ that converges in $C(X)$. This is a contradiction and $\mathcal{F}$ is consequently bounded. Fix $\epsilon > 0$. Since $\mathcal{F}$ is relatively compact, there is a finite collection of n open balls of radius $\frac{\epsilon}{3}$ that cover $\mathcal{F}$:

$$\mathcal{F} \subseteq \bigcup_{i=1}^{n} B_{\frac{\epsilon}{3}}(f_k).$$

The finite collection $\{f_1, \ldots, f_n\}$ is uniformly equicontinuous. There is a $\delta > 0$ such that

$$d(x, y) < \delta \quad \Longrightarrow \quad |f_j(x) - f_j(y)| < \frac{\epsilon}{3}$$

for all $j = 1, \ldots, n$. Therefore, for any $f \in \mathcal{F}$ we have

$$|f(x) - f(y)| \leq |f(x) - f_k(x)| + |f_k(x) - f_k(y)| + |f_k(y) - f(y)| \leq \epsilon$$

whenever $d(x, y) < \delta$. Thus, $\mathcal{F}$ is equicontinuous.

Suppose now that $\mathcal{F}$ is bounded and equicontinuous. Since X is compact and metric, it is separable. Let $Q = \{x_k\}_{k \in \mathbb{N}}$ be a countable set dense in X. Since $\mathcal{F}$ is bounded, the set of real numbers

$$\|f_n(x_k)\| < \mathcal{C}$$

is bounded for any $f_n \in \mathcal{F}$ and $x_k \in Q$. Since $\mathbb{R}$ is a compact metric space, for each $k \in \mathbb{N}$ the sequence

$$\{f_n(x_k)\}_{n \in \mathbb{N}}$$

contains a convergent subsequence $\{f_{n_j}(x_k)\}_{n_j \in \mathbb{N}}$. By Theorem 1.3.7 the subsequence $\{f_{n_j}\}_{n_j \in \mathbb{N}}$ converges in $C(X)$ and $\mathcal{F}$ is precompact. $\qquad\square$

As noted earlier, many of the most common topological spaces employed by engineers and scientists in applications are in fact metric spaces. The utility of these spaces provides evidence of the power of abstract methods to unify diverse concepts in applications. We now consider a collection of examples that illustrate the generality of these ideas. The first two examples are quite elementary.

Example 1.3.2. *The simplest useful metric space is, perhaps, the set of real numbers* $\mathbb{R}$ *endowed with the usual distance function*

$$d : \mathbb{R} \times \mathbb{R} \longrightarrow [0, \infty)$$
$$d(x, y) \triangleq |x - y|.$$

Clearly, this choice of d defines a real-valued, finite and nonnegative function on $\mathbb{R} \times \mathbb{R}$. *The function d is symmetric* (M3), *satisfies the triangle inequality* (M4) *and vanishes if and only if* $x = y$ (M2).

Example 1.3.3. *It is likewise fundamental that the collection of all complex numbers* $\mathbb{C}$, *endowed with the distance function that evaluates the magnitude of the difference*

$$d : \mathbb{C} \times \mathbb{C} \longrightarrow [0, \infty)$$
$$d(z_1, z_2) \triangleq |z_1 - z_2|$$

is a metric space. Properties (M1), (M2), (M3) *and* (M4) *are all easy to verify.*

Passing to more general examples can make it increasingly difficult to verify the hypotheses (M1)–(M4). The following example, which is a simple extension of the previous cases in Examples 1.3.2 and 1.3.3, illustrates this fact.

Example 1.3.4. *Let either $\mathbb{R}^N$, or $\mathbb{C}^N$, denote the set of N-tuples $\{x_1, x_2, x_3, \ldots, x_N\}$ each of whose entries x_i are extracted either from $\mathbb{R}$, or $\mathbb{C}$, respectively. It turns out that there are many interesting and useful metrics that can be defined on $\mathbb{R}^N$, or $\mathbb{C}^N$. The most common of these metrics are given by*

$$d_p(x, y) \triangleq \left(\sum_{i=1}^{N} |x_i - y_i|^p \right)^{\frac{1}{p}}.$$

We claim that each function d_p is a metric for $1 \le p < \infty$. It is trivial to establish (M1), (M2) and (M3). However, the verification of (M4), the triangle inequality, entails considerably more work. To establish this fact, we will need to derive two well-known inequalities. These inequalities are the subject of the next section.

1.3.8 Hölder's and Minkowski's Inequalities

In this section we will derive two of the most commonly employed inequalities in analysis: Hölder's and Minkowski's inequalities. Their derivation is not difficult, it is simply lengthy. However, the derivation does illustrate the details that must be considered in proving that functions are indeed metrics. The proof of Hölder's and Minkowski's inequalities rely on still another, more fundamental inequality: Young's inequality.

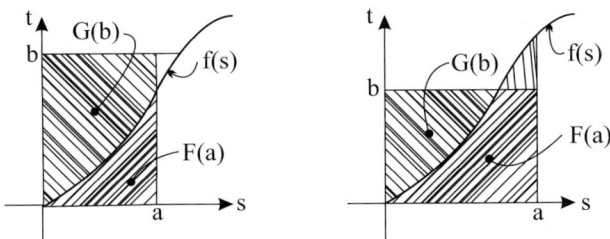

Figure 1.10: Young's Inequality

Proposition 1.3.1. *Let f be a real-valued, continuous, strictly increasing function defined on $[0, \infty)$ such that*

$$\lim_{x \to \infty} f(x) = \infty, \qquad f(0) = 0.$$

Let $g \triangleq f^{-1}$. Define the integrals

$$F(x) = \int_0^x f(s)ds, \qquad G(x) = \int_0^x g(t)dt.$$

Then if a and b are nonnegative and finite, we have

$$ab \le F(a) + G(b).$$

Proof. This proof is immediate when we consider the Figure 1.10. In either case, we see that the area ab is less than or equal to the area of the sum of the shaded regions $F(a)$ and $G(b)$. $\qquad\square$

Corollary 1.3.1. *For $p > 1$, $\frac{1}{p} + \frac{1}{q} = 1$ and a, b nonnegative and finite, we have*

$$ab \leq \frac{a^p}{p} + \frac{b^q}{q}.$$

Proof. Suppose we choose $f(s) = s^{p-1}$ so that

$$g(t) = t^{\frac{1}{p-1}}.$$

Then

$$F(a) = \int_0^a s^{p-1} ds = \frac{1}{p} a^p$$

and

$$G(b) = \int_0^b t^{\frac{1}{p-1}} dt = \left(\frac{1}{p-1} + 1\right)^{-1} b^{\frac{1}{p-1}+1} = \frac{p-1}{p} b^{\frac{p}{p-1}} = \frac{1}{q} b^q.$$

Then Young's inequality guarantees that

$$ab \leq \frac{1}{p} a^p + \frac{1}{q} b^q$$

and the corollary is proved. $\qquad\square$

Theorem 1.3.9 (Hölder's Inequality). *Let p, q be two integers such that $1 < p < \infty$, $\frac{1}{p} + \frac{1}{q} = 1$. Then Hölder's inequality holds*

$$\sum_{i=1}^N |x_i y_i| \leq \left(\sum_{i=1}^N |x_i|^p\right)^{\frac{1}{p}} \left(\sum_{i=1}^N |y_i|^q\right)^{\frac{1}{q}}.$$

Proof. We have shown that

$$ab \leq \frac{1}{p} a^p + \frac{1}{q} b^q$$

choose

$$a_k = \frac{|x_k|}{\left(\sum_{k=1}^N |x_k|^p\right)^{\frac{1}{p}}}, \qquad b_k = \frac{|y_k|}{\left(\sum_{k=1}^N |y_k|^q\right)^{\frac{1}{q}}}.$$

Then we have

$$\sum_{i=1}^N \frac{|x_i y_i|}{\left(\sum_{i=1}^N |x_k|^p\right)^{\frac{1}{p}} \left(\sum_{i=1}^N |y_k|^q\right)^{\frac{1}{q}}} \leq \sum_{i=1}^N \left\{ \frac{1}{p} \frac{|x_i|^p}{\sum_{k=1}^N |x_k|^p} + \frac{1}{q} \frac{|y_i|^q}{\sum_{k=1}^N |y_k|^q} \right\}$$

$$= \frac{1}{p} \frac{\sum_{i=1}^N |x_i|^p}{\sum_{k=1}^N |x_k|^p} + \frac{1}{q} \frac{\sum_{i=1}^N |y_i|^q}{\sum_{k=1}^N |y_k|^q} = \frac{1}{p} + \frac{1}{q} = 1.$$

The theorem is proved. $\qquad\square$

Theorem 1.3.10 (Minkowski's Inequality). *Let p be an integer such that $1 \leq p < \infty$. Then Minkowski's inequality holds*

$$\left(\sum_{i=1}^{N} |x_i \pm y_i|^p \right)^{\frac{1}{p}} \leq \left(\sum_{i=1}^{N} |x_i|^p \right)^{\frac{1}{p}} + \left(\sum_{i=1}^{N} |y_i|^p \right)^{\frac{1}{p}}.$$

Proof. Consider

$$\sum_{i=1}^{N} |x_i + y_i|^p \leq \sum_{i=1}^{N} (|x_i| + |y_i|)^p$$

$$= \sum_{i=1}^{N} (|x_i| + |y_i|)^{p-1} |x_i| + \sum_{i=1}^{N} (|x_i| + |y_i|)^{p-1} |y_i|.$$

We can now apply Hölder's inequality to each term on the right-hand side of the above inequality

$$\sum_{i=1}^{N} (|x_i| + |y_i|)^{p-1} |x_i| \leq \left(\sum_{i=1}^{N} (|x_i| + |y_i|)^{(p-1)q} \right)^{\frac{1}{q}} \left(\sum_{i=1}^{N} |x_i|^p \right)^{\frac{1}{p}}$$

$$\sum_{i=1}^{N} (|x_i| + |y_i|)^{p-1} |y_i| \leq \left(\sum_{i=1}^{N} (|x_i| + |y_i|)^{(p-1)q} \right)^{\frac{1}{q}} \left(\sum_{i=1}^{N} |y_i|^p \right)^{\frac{1}{p}}$$

and obtain

$$\sum_{i=1}^{N} (|x_i| + |y_i|)^p \leq \left(\sum_{i=1}^{N} (|x_i| + |y_i|)^{(p-1)q} \right)^{\frac{1}{q}} \left\{ \left(\sum_{i=1}^{N} |x_i|^p \right)^{\frac{1}{p}} + \left(\sum_{i=1}^{N} |y_i|^p \right)^{\frac{1}{p}} \right\}$$

$$= \left(\sum_{i=1}^{N} (|x_i| + |y_i|)^p \right)^{\frac{1}{q}} \left\{ \left(\sum_{i=1}^{N} |x_i|^p \right)^{\frac{1}{p}} + \left(\sum_{i=1}^{N} |y_i|^p \right)^{\frac{1}{p}} \right\}.$$

Dividing right- and left-hand sides of this inequality by $\left(\sum_{i=1}^{N} (|x_i| + |y_i|)^p \right)^{\frac{1}{q}}$, we obtain

$$\left(\sum_{i=1}^{N} (|x_i| + |y_i|)^p \right)^{1-\frac{1}{q}} \leq \left(\sum_{i=1}^{N} |x_i|^p \right)^{\frac{1}{p}} + \left(\sum_{i=1}^{N} |y_i|^p \right)^{\frac{1}{p}}$$

so that

$$\left(\sum_{i=1}^{N} |x_i + y_i|^p \right)^{\frac{1}{p}} \leq \left(\sum_{i=1}^{N} |x_i|^p \right)^{\frac{1}{p}} + \left(\sum_{i=1}^{N} |y_i|^p \right)^{\frac{1}{p}}. \qquad \square$$

Corollary 1.3.2. *Let $\{x_i\}_{i \in \mathbb{N}}$ and $\{y_i\}_{i \in \mathbb{N}}$ be such that*

$$\left(\sum_{i=1}^{\infty} |x_i|^p \right)^{\frac{1}{p}} < \infty, \qquad \left(\sum_{i=1}^{\infty} |y_i|^q \right)^{\frac{1}{q}} < \infty$$

for $1 \leq q < \infty$, $1 \leq p < \infty$, and $\frac{1}{p} + \frac{1}{q} = 1$. Then, Hölder's and Minkowski's inequalities hold, respectively

$$\sum_{i=1}^{\infty} |x_i y_i| \leq \left(\sum_{i=1}^{\infty} |x_i|^p \right)^{\frac{1}{p}} \left(\sum_{i=1}^{\infty} |y_i|^q \right)^{\frac{1}{q}}$$

$$\left(\sum_{i=1}^{\infty} |x_i \pm y_i|^p \right)^{\frac{1}{p}} \leq \left(\sum_{i=1}^{\infty} |x_i|^p \right)^{\frac{1}{p}} + \left(\sum_{i=1}^{\infty} |y_i|^p \right)^{\frac{1}{p}}.$$

Proof. As in the proof of the finite-dimensional Hölder's inequality, we know that

$$ab \leq \frac{1}{p} a^p + \frac{1}{q} b^q.$$

We can choose

$$a_k \triangleq \frac{|x_k|}{\left(\sum_{k=1}^{\infty} |x_k|^p \right)^{\frac{1}{p}}}, \qquad b_k \triangleq \frac{|y_k|}{\left(\sum_{k=1}^{\infty} |y_k|^q \right)^{\frac{1}{q}}}.$$

In this case, the denominator in each expression makes sense by hypothesis. Following the same line of reasoning as in the derivation of Theorem 1.3.9, we have

$$\sum_{i=1}^{N} \frac{|x_i y_i|}{\left(\sum_{k=1}^{\infty} |x_k|^p \right)^{\frac{1}{p}} \left(\sum_{k=1}^{\infty} |y_k|^q \right)^{\frac{1}{q}}} \leq \frac{1}{p} + \frac{1}{q} = 1$$

or

$$\sum_{i=1}^{N} |x_i y_i| \leq \left(\sum_{k=1}^{\infty} |x_k|^p \right)^{\frac{1}{p}} \left(\sum_{k=1}^{\infty} |y_k|^q \right)^{\frac{1}{q}}.$$

Again, the right-hand side makes sense by hypothesis. We can take the limit on the left-hand side to obtain Hölder's inequality

$$\sum_{i=1}^{\infty} |x_i y_i| \leq \left(\sum_{k=1}^{\infty} |x_k|^p \right)^{\frac{1}{p}} \left(\sum_{k=1}^{\infty} |y_k|^q \right)^{\frac{1}{q}}.$$

By similar reasoning, we can follow the proof in Theorem 1.3.10 to obtain Minkowski's inequality for infinite sequences. □

We can now return to Example 1.3.4 and show that the function d_p does satisfy a triangle inequality.

Example 1.3.5. *Now, with Hölder's and Minkowski's inequalities, it is simple to show that $\mathbb{R}^N$ or $\mathbb{C}^N$ is a metric space when equipped with the metric*

$$d_p(x,y) \triangleq \left(\sum_{i=1}^{N} |x_i - y_i|^p \right)^{\frac{1}{p}}.$$

Clearly the function $d_p(\cdot,\cdot)$ is symmetric, real-valued, finite and nonnegative. The only property remaining to be shown is that $d_p(\cdot,\cdot)$ satisfies the triangle inequality. Suppose $x, y, z \in \mathbb{R}^N$. We can calculate

$$d_p(x,z) = \left(\sum_{i=1}^{N} |x_i - z_i|^p \right)^{\frac{1}{p}}$$

$$= \left(\sum_{i=1}^{N} |(x_i - y_i) + (y_i - z_i)|^p \right)^{\frac{1}{p}}$$

$$\leq \left(\sum_{i=1}^{N} |x_i - y_i|^p \right)^{\frac{1}{p}} + \left(\sum_{i=1}^{N} |y_i - z_i|^p \right)^{\frac{1}{p}}$$

$$= d_p(x,y) + d_p(y,z).$$

The function $d_p(\cdot,\cdot)$ is a metric.

Thus far, all of our examples have dealt with collections, or sequences, of real or complex numbers. However, the strength of abstraction is that we can treat diverse mathematical entities. The next example shows that many collections of functions can be cast as metric spaces with the definition of an appropriate metric function.

Example 1.3.6. *One common metric space studied in this book is constructed as follows. We take the collection of all real-valued functions that are continuous on the closed interval $[a,b]$. We define a candidate for a metric on this set by the function*

$$d(f,g) \triangleq \max_{x \in [a,b]} |f(x) - g(x)|.$$

This collection of sets, equipped with the above distance function defines a metric space $C[a,b]$. Since f, g are continuous on $[a,b]$, their sum is continuous on $[a,b]$. Since every continuous function defined over a closed and bounded set attains its maximum (the Weierstrass Theorem), the function d is well defined for any pair of continuous functions. Moreover, it is clear that d is real-valued, finite and nonnegative. Symmetry of the function d is obvious

$$d(f,g) = d(g,f)$$

as is the property

$$d(f,g) = 0 \qquad \Longleftrightarrow \qquad f = g.$$

To show that a triangle inequality holds for the function d, let f, g and h be continuous functions on $[a, b]$. For each $x \in [a, b]$ we can write

$$|f(x) - h(x)| \leq |f(x) - g(x)| + |g(x) - h(x)|$$

by the triangle inequality for the real numbers $f(x)$, $g(x)$ and $h(x)$. We can take the maximum over both sides of the above inequality to obtain

$$\max_{x \in [a,b]} |f(x) - h(x)| \leq \max_{x \in [a,b]} \left\{ |f(x) - g(x)| + |g(x) - h(x)| \right\}$$

$$\leq \max_{x \in [a,b]} |f(x) - g(x)| + \max_{x \in [a,b]} |g(x) - h(x)|.$$

But this means that

$$d(f, h) \leq d(f, g) + d(g, h)$$

and the function $d(f, g)$ is a metric.

In Example 1.3.4 we showed that N-tuples of real or complex functions can be made a metric space with the choice of many different metrics. In the following example, we see that the same is true of infinite sequences of real or complex numbers. Again, the crucial step is establishing the triangle inequality via Hölder's and Minkowski's inequalities.

Example 1.3.7. The Space l_p. *The metric spaces $(\mathbb{R}^N, d_p)$ and $(\mathbb{C}^N, d_p)$ are defined in terms of finite sequences. Extensions to metric spaces defined in terms of infinite sequences can be constructed in an analogous fashion. We define the space l_p on $\mathbb{R}$ or $\mathbb{C}$ to be the collection of all (infinite) sequences in $\mathbb{R}$ or $\mathbb{C}$ such that*

$$\left(\sum_{i=1}^{\infty} |x_i|^p \right)^{\frac{1}{p}} < \infty.$$

In other words,

$$l_p = \left\{ \{x_i\}_{i \in \mathbb{N}} \subset \mathbb{R} : \left(\sum_{i=1}^{\infty} |x_i|^p \right)^{\frac{1}{p}} < \infty \right\}$$

or

$$l_p = \left\{ \{x_i\}_{i \in \mathbb{N}} \subset \mathbb{C} : \left(\sum_{i=1}^{\infty} |x_i|^p \right)^{\frac{1}{p}} < \infty \right\}.$$

To show that these spaces are metric spaces, we use Corollary 1.3.2 establishing infinite-dimensional versions of Hölder's and Minkowski's inequalities. With these two inequalities it is now straightforward to show that l_p for $1 \leq p \leq \infty$ is a metric space when equipped with the distance function.

$$d_p(x, y) = \left(\sum_{i=1}^{\infty} |x_i - y_i|^p \right)^{\frac{1}{p}}.$$

First it is clear that $d_p(x, y)$ is real-valued and finite for all $\{x_k\}_{k\in\mathbb{N}} \in l_p$ and $\{y_k\}_{k\in\mathbb{N}} \in l_p$. Minkowski's inequality guarantees that the metric is finite.

$$d_p(x, y) = \left(\sum_{i=1}^{\infty} |x_i - y_i|^p \right)^{\frac{1}{p}} \leq \left(\sum_{i=1}^{\infty} |x_i|^p \right)^{\frac{1}{p}} + \left(\sum_{i=1}^{\infty} |y_i|^p \right)^{\frac{1}{2}} < \infty.$$

Finally, we also see that the function $d_p(\cdot, \cdot)$ satisfies the triangle inequality. Suppose $x, y, z \in l_p$. We have

$$d_p(x, z) = \left(\sum_{i=1}^{\infty} |x_i - y_i + y_i - z_i|^p \right)^{\frac{1}{p}}$$

$$\leq \left(\sum_{i=1}^{\infty} |x_i - y_i|^p \right)^{\frac{1}{p}} + \left(\sum_{i=1}^{\infty} |y_i - z_i|^p \right)^{\frac{1}{p}}$$

by Minkowski's inequality. But this simply says that

$$d_p(x, y) \leq d(x, y) + d(y, z)$$

for any $x, y, z \in l_p$.

1.4 Vector Spaces

Until now, we have not introduced any *algebraic structure* in our considerations of abstract spaces. Of course, there are many interesting fields of study that comprise modern abstract algebra. In fact, group theory and the study of rings are two such fields that have had a profound impact in areas related to mechanics just over the past few years. For example [22] studies group theoretic methods for the identification of invariants of dynamical systems that govern physical systems. The study of rings and fields plays an important role in control theory [36].

We will deal with only one class of algebraic structure in this book: the linear, or vector, space. As usual, when viewed axiomatically, a linear space can be rigorously defined in terms of an Abelian group (the vectors) and a field (the scalars), and a collection of allowable rules for combining these two species of mathematical entities. See for examples, [21, 17]. For our purposes, much less detail will suffice.

Definition 1.4.1. *A set that is closed under the two algebraic operations of vector addition and multiplication by scalars is called a vector space.*

Again, we emphasize that while this definition is not "axiomatic", it does convey the essential ingredients of a vector space. The nature of this definition is made clear if we consider a few simple examples.

Example 1.4.1. *In Example* (1.3.4), *we saw that the set of N-tuples $\{x_1, \ldots, x_N\} \subseteq \mathbb{R}^N$ or $\mathbb{C}^N$ comprised a metric space. Define the addition of vectors $x, y \in \mathbb{R}^N$ or $\mathbb{C}^N$ in the obvious way*

$$x + y = z \quad \Longleftrightarrow \quad \left\{\begin{matrix} x_1 \\ \vdots \\ x_N \end{matrix}\right\} + \left\{\begin{matrix} y_1 \\ \vdots \\ y_N \end{matrix}\right\} = \left\{\begin{matrix} x_1 + y_1 \\ \vdots \\ x_N + y_N \end{matrix}\right\} = \left\{\begin{matrix} z_1 \\ \vdots \\ z_N \end{matrix}\right\}.$$

Likewise, define the multiplication of a vector $x \in \mathbb{R}^N$ or $\mathbb{C}^N$ as

$$\alpha x = y \quad \Longleftrightarrow \quad \alpha \left\{\begin{matrix} x_1 \\ \vdots \\ x_N \end{matrix}\right\} = \left\{\begin{matrix} \alpha x_1 \\ \vdots \\ \alpha x_N \end{matrix}\right\} = \left\{\begin{matrix} y_1 \\ \vdots \\ y_N \end{matrix}\right\}.$$

In this case it is clear that $\mathbb{R}^N$ or $\mathbb{C}^N$ constitute a vector space: they are closed under addition of vectors and multiplication of vectors by scalars. The reader should note that the considerations above are purely algebraic. It does not matter, in this example, what metric d_p from Example 1.3.4 is defined on $\mathbb{R}^N$ or $\mathbb{C}^N$.

Perhaps it is no surprise to the reader that our common notion of vectors or N-tuples define a vector space. Again, we emphasize that the general definition of a vector space encompasses many mathematical objects, as the next example illustrates.

Example 1.4.2. *In Example 1.3.6, we showed that the set of all continuous, real-valued functions $C[a, b]$ defined on the closed interval $[a, b]$ is a metric space when endowed with the uniform metric*

$$d(f, g) \triangleq \max_{x \in [a,b]} |f(x) - g(x)|.$$

It is also not difficult to see that $C[a, b]$ is a vector space. We define the addition of two vectors $f, g \in C[a, b]$ in the expected way

$$h = f + g \quad \Longleftrightarrow \quad h(x) = f(x) + g(x) \qquad \forall x \in [a, b]. \tag{1.5}$$

The definition of multiplication of a vector by a scalar is also intuitive:

$$g = \alpha f \quad \Longleftrightarrow \quad g(x) = \alpha f(x) \qquad \forall x \in [a, b]. \tag{1.6}$$

Again, it is clear that the addition of two continuous, real-valued functions $f, g \in C[a, b]$ generate a function $h \in C[a, b]$ according to the definition in Equation (1.5). The multiplication of a function $f \in C[a, b]$ by a real number $\alpha \in \mathbb{R}$ in Equation (1.6) generates a continuous function $g \in C[a, b]$. Hence, $C[a, b]$ is closed under the defined operators of vector addition and scalar multiplication. It is a vector space.

In fact, most of the examples considered in our discussion of metric spaces are also examples of vector spaces. The next example shows that the metric space l_p can also be construed as a vector space, with the appropriate definitions of vector addition and scalar multiplication.

Example 1.4.3. *Consider the sequence spaces l_p defined on $\mathbb{R}$ or $\mathbb{C}$ discussed in Example 1.3.7. Addition of two vectors $f, g \in l_p$ and multiplication of a vector by a scalar are defined as*

$$
\begin{aligned}
h = f + g &\iff \{h_i\}_{i\in\mathbb{N}} = \{f_i + g_i\}_{i\in\mathbb{N}} \\
g = \alpha f &\iff \{g_i\}_{i\in\mathbb{N}} = \{\alpha f_i\}_{i\in\mathbb{N}}.
\end{aligned}
\tag{1.7}
$$

Clearly, the algebraic manipulations make sense on a component-wise basis. To show that the set l_p is closed under vector addition and multiplication of vectors by scalars, we must show that $h \in l_p$ and $g \in l_p$ in (1.7). From Minkowski's Inequality we can write

$$
\left(\sum_{i\in\mathbb{N}} |h_i|^p\right)^{\frac{1}{p}} \le \left(\sum_{i\in\mathbb{N}} |f_i|^p\right)^{\frac{1}{p}} + \left(\sum_{i\in\mathbb{N}} |g_i|^p\right)^{\frac{1}{p}}.
$$

Since $f, g \in l_p$ by hypothesis, the terms on the right of this inequality are finite. Hence, the summation on the left of the inequality is finite, and $h \in l_p$. Finally, we can directly compute

$$
\left(\sum_{i\in\mathbb{N}} |\alpha f_i|^p\right)^{\frac{1}{p}} = |\alpha| \left(\sum_{i\in\mathbb{N}} |f_i|^p\right)^{\frac{1}{p}} < \infty
$$

and $g \in l_p$.

In our discussions of topological spaces, we learned that the characterization of the topology is often carried out by defining the building blocks, or bases and local bases, of the topology. For vector spaces, it is again essential to identify building blocks for the vector spaces under consideration. The bases for vector spaces are defined in terms of algebraic notions of linear independence of elements of a vector space.

Definition 1.4.2. *Let X be a vector space. A linear combination of a finite collection of n vectors $x_1, \ldots, x_n$ contained in X is any vector y that can be written as*

$$
y = \sum_{i=1}^{n} \alpha_i x_i
$$

for some constants $\alpha_i \in \mathbb{R}$ or $\mathbb{C}$, $i = 1, \ldots, n$.

Definition 1.4.3. *Let X be a vector space. A finite set of n vectors $x_1, \ldots, x_n$ contained in X is said to be linearly independent if none of the vectors can be* *written as a linear combination of the remaining vectors. Alternatively, the set* $\{x_1, \ldots, x_n\} \subseteq X$ *is linearly independent if the sum*

$$\alpha_1 x_1 + \alpha_2 x_2 + \ldots + \alpha_n x_n = 0$$

holds only for the choice of constants

$$\alpha_1 = \alpha_2 = \ldots = \alpha_n = 0.$$

With the definition of a linearly independent set of vectors, we can define a basis for a vector space.

Definition 1.4.4. *Let X be a vector space and suppose S is a set of linearly independent vectors contained in X. The set S is a basis for X if every element of X can be written uniquely as a finite linear combination of elements in S. The number of elements in S is the dimension of the vector space X.*

For some simple cases, the basis for a vector space is self-evident. For example, if X is a set and $\{x_1, \ldots, x_n\}$ is a finite collection of linearly independent elements in X, we define the span $\{x_1, \ldots, x_n\}$ to be

$$\text{span } \{x_i\}_{i=1}^n = \Big\{ x \in X : x = \sum_{i=1}^n \alpha_i x_i, \ \alpha_i \in \mathbb{R} \Big\}.$$

By construction, the span of a set of linearly independent vectors is a vector space and the generating set $\{x_i\}_{i=1}^n$ is a basis for this space. In other cases, it is not nearly so evident that we can construct a basis for a given vector space. To show that every vector space has a basis, we require an abstract theorem. One of the foundations of functional analysis is Zorn's lemma, which can be employed to prove many other fundamental theorems, including, for example, the Hahn-Banach theorem.

Lemma 1.4.1 (Zorn's Lemma). *Let X be a nonempty partially ordered set. Suppose every totally ordered subset of X has an upper bound. Then X has a maximal element.*

Like many of the essential, fundamental results of functional analysis, the impact of this lemma is probably not clear at first glance. We briefly present an application of Zorn's lemma. We will see in many applications in this book that there is a similarity in the structure of the proofs based on Zorn's lemma. That is, there is a common framework for application of the lemma.

Proposition 1.4.1. *Let X be a nontrivial vector space. Then X has a basis.*

Proof. We construct our partially ordered set Y as the collection of all subsets of linearly independent elements extracted from X. That is,

$$\mathcal{Y} \in Y \iff \mathcal{Y} = \{y_1, \ldots, y_n\}$$

for some $n \in \mathbb{N}$, where $y_i \in X$, $i = 1, \ldots, n$, are linearly independent. The set $\mathcal{Y}$ is partially ordered by set inclusion

$$\mathcal{Y} \leq \mathcal{Z} \iff \mathcal{Y} = \{y_1, \ldots, y_n\} \subseteq \mathcal{Z} = \{z_1, \ldots, z_m\}$$

for some $m \geq n$. If $\{\mathcal{Y}_\alpha\}_{\alpha \in A}$ is a totally ordered subset of Y, then $\{\mathcal{Y}_\alpha\}_{\alpha \in A}$ has an upper bound. The upper bound is just the union of all $\mathcal{Y}_\alpha$ in the totally ordered subset $\{\mathcal{Y}_\alpha\}_{\alpha \in A}$

$$\mathcal{Y}_u = \bigcup_\alpha \mathcal{Y}_\alpha.$$

According to Zorn's lemma, there is a maximal element $\mathcal{Y}_m$ in Y. We claim that $\mathcal{Y}_m$ is a basis for X. Suppose to the contrary that these exists some $x_0 \in X$ that cannot be expressed as a linear combination of elements in $\mathcal{Y}_m$. Then the set $\mathcal{Y}_m \cup \{x_0\}$ is linearly independent and contains $\mathcal{Y}_m$. This contradicts the maximality of $\mathcal{Y}_m$ and the proposition is proven. $\qquad\square$

1.5 Normed Vector Spaces

Once a vector space is defined, it is possible to endow it with a topology. The most common method for defining a topology on a vector space is to specify a length for each vector. The function that associates a length to each vector is called the *norm* on a vector space.

Definition 1.5.1. *A normed vector space X is a vector space with a real-valued function $\| \cdot \|$ with the properties that for all $x, y \in X$ and $\alpha \in \mathbb{R}$ (or $\mathbb{C}$)*

(N1) $\|x\| \geq 0$

(N2) $\|x\| = 0 \iff x = 0$

(N3) $\|\alpha x\| = |\alpha| \|x\|$

(N4) $\|x + y\| \leq \|x\| + \|y\|$.

Every normed vector space is a metric space, and consequently, a topological space. We can always define a metric $d(\cdot, \cdot)$ on a normed vector space in terms of its norm

$$d(x, y) \equiv \|x - y\| \quad \forall x, y \in X.$$

1.5.1 Basic Definitions

By recalling the notion of convergence in a topological space, and interpreting this definition with the additional structure of the normed vector space, we can define convergence in a normed vector space.

Definition 1.5.2. *A sequence $\{x\}_{n \in \mathbb{N}}$ in a normed vector space X is said to be convergent if $\exists x \in X$ such that*

$$\lim_{n \to \infty} \|x_n - x\| = 0.$$

Likewise, we can define a Cauchy sequence via the definition introduced for general metric spaces.

Definition 1.5.3. *A sequence $\{x_n\}_{n \in \mathbb{N}}$ in a normed vector space X is Cauchy if for every $\epsilon > 0$ there is an $N(\epsilon)$ such that*

$$\|x_m - x_n\| < \epsilon \quad \forall m, n > N(\epsilon).$$

Hence, a normed vector space X is complete if the norm defined on X induces a metric for which X is a complete metric space. A complete normed vector space is the foundation on which we define most abstract spaces that are of practical use to engineers and scientists.

Definition 1.5.4. *A complete normed space is called a Banach space.*

Again, as we noted for the metric space, there are useful alternative characterizations of a continuous function, a closed set and a compact set in a normed vector space.

Definition 1.5.5. *Let (X, Y) be normed vector spaces and let $f : X \to Y$. The function f is continuous at $x_0 \in X$ if for each $\epsilon > 0$, $\exists \delta(\epsilon) > 0$ such that*

$$\|x - x_0\|_X \leq \delta(\epsilon) \implies \|f(x) - f(x_0)\|_Y < \epsilon.$$

Definition 1.5.6. *Let X be a normed vector space. A subset $M \subseteq X$ is closed if and only if it contains all of its accumulation points.*

Definition 1.5.7. *Let X be a normed vector space. A subset $M \subseteq X$ is compact if and only if every sequence extracted from M contains a subsequence that converges to an element of M.*

We leave to the reader the exercise of demonstrating that these definitions of a closed set and a compact set agree with the definition for closed and compact sets in a topological space.

1.5.2 Examples of Normed Vector Spaces

As in previous sections, we begin our discussion of examples by studying the most elementary cases.

Example 1.5.1. *Again, a prototypical example of a normed vector space is simply the vector space of real numbers $\mathbb{R}$ where we endow $\mathbb{R}$ with the norm*

$$\|x\| \triangleq |x|.$$

Properties (N1) through (N3) follow simply from the definition of the absolute value function, while (N4) is the celebrated triangle inequality for real numbers.

The vector spaces of N-tuples of real or complex numbers are examples of normed vector spaces also. The norm is obvious and can be deduced from our discussion of metrics on the set of real numbers.

Example 1.5.2. *We can endow the vector spaces of N-tuples of real numbers, $\mathbb{R}^n$, or N-tuples of complex numbers $\mathbb{C}^n$, with many different norms. We define the p-norm*

$$\|x\|_p = \left(\sum_{i=1}^{n} |x_i|^p \right)^{\frac{1}{p}} \tag{1.8}$$

for $x \in \mathbb{R}^n$ or $\mathbb{C}^n$. These normed vector spaces are denoted by $\left(\mathbb{R}^n, \|\cdot\|_p\right)$ and $\left(\mathbb{C}^n, \|\cdot\|_p\right)$, respectively. That these functions are in fact norms over $\mathbb{R}^n$ and $\mathbb{C}^n$ is not difficult to establish. Clearly, properties (N1) through (N3) are an immediate consequence of Definition 1.8. The triangle inequality expressed in property (N4) follows from Minkowski's inequality in Theorem 1.3.10.

And, once again, normed vector spaces can be constructed from rather abstract objects, ones that might not immediately come to mind. The following discussion presents one such example.

Example 1.5.3. *Example 1.4.2 showed that the collection of all continuous functions $C[a, b]$ is a vector space. The vector space $C[a, b]$ is frequently endowed with a norm*

$$\|f\|_\infty = \max_{x \in [a,b]} |f(x)|. \tag{1.9}$$

As in all of our previous cases, properties (N1), (N2), and (N3) follow directly from the definition of the norm in Equation (1.9). Property (N4), the triangle inequality follows from the proof of the triangle inequality for the metric on $C[a, b]$ in Example 1.3.6.

We saw in Example 1.5.2 that N-tuples of real or complex numbers constitute a normed vector space, as our intuition would suggest. Likewise, the next example shows that infinite sequences of real or complex numbers can be construed as normed vector spaces. Again, the definition of the norm is critical and is related to the metric discussed in Example 1.4.3.

Example 1.5.4. *Consider the collection of all sequences $\{x_k\}_{k\in\mathbb{N}} \subseteq \mathbb{R}$ or $\{x_k\}_{k\in\mathbb{N}} \subseteq \mathbb{C}$ such that the expression*

$$\|x\|_p \triangleq \left\{ \sum_{k=1}^{\infty} |x_k|^p \right\}^{\frac{1}{p}}$$

is finite. The sets

$$l_p \triangleq \left\{ \{x_k\}_{k\in\mathbb{N}} \subseteq \mathbb{R} : \|x\|_p < \infty \right\}$$

and

$$l_p \triangleq \left\{ \{x_k\}_{k\in\mathbb{N}} \subseteq \mathbb{C} : \|x\|_p < \infty \right\}$$

are normed vector spaces. As in all of our previous examples, it is clear that

(N1) $\|x\|_p \geq 0$

(N2) $\|x\|_p = 0 \iff x = 0$

(N3) $\|\alpha x\|_p = |\alpha| \, \|x_p\|.$

By Minkowski's inequality for infinite sums, we also can conclude that

$$\|x + y\|_p \leq \|x\|_p + \|y\|_p \tag{1.10}$$

whenever x, $y \in l_p$. Finally, it is clear in view of the above considerations that the set l_p is in fact a vector space. Suppose $x \in l_p$. Then

$$\alpha x = \{\alpha x_i\}_{i \in \mathbb{N}}$$

and

$$\|\alpha x\|_p = \left(\sum_{i=1}^{\infty} |\alpha x_i|^p \right)^{\frac{1}{p}}$$
$$= \sum_{i=1}^{\infty} \left(|\alpha|^p |x_i|^p \right)^{\frac{1}{p}}$$
$$= |\alpha| \|x\|_p < \infty.$$

Hence $\alpha x \in l_p$ and l_p is closed with respect to multiplication by scalars $\alpha \in \mathbb{R}$. Similarly, suppose x, $y \in l_p$. Then by the triangle inequality in Equation (1.10) the norm $\|x + y\|_p$ is finite, and the set l_p is closed with respect to addition of vectors.

In the next example, we consider a norm that has not appeared previously in this text. While it is somewhat less frequently encountered than the previous examples, it plays a crucial role in some applications.

Example 1.5.5. *Let f be a real-valued function on an interval $[a, b] \subset \mathbb{R}$. A partition P on $[a, b]$ is a finite set $\{x_k\}_{k=0}^{n}$ such that*

$$a = x_0 < x_1 < x_2 < \cdots < x_n = b.$$

The variation of the function f over a partition P is defined to be

$$V(f; P) \triangleq \sum_{i=0}^{n-1} |f(x_{i+1}) - f(x_i)|.$$

The total variation of a function is defined to be the supremum of the variation over all possible partitions. Suppose $\mathcal{P}$ is the collection of all partitions defined on the interval $[a, b]$. Then the total variation of f is given by

$$V(f) \triangleq \sup_{P \in \mathcal{P}} \sum_{i=0}^{n-1} |f(x_{i+1}) - f(x_i)|.$$

*A function is defined to be a function of bounded variations provided $V(f) < \infty$.
The set of all functions of bounded variation defined on the interval $[a, b]$ is denoted
$BV[a, b]$. The variation of a function $V(f)$ is equal to zero if and only if the
function f is identically equal to a constant on $[a, b]$. Thus, $V(\cdot)$ is not a norm
on $BV[a, b]$. It is simple to modify it, however, to define a norm on $BV[a, b]$. For
example, the expression*

$$\|f\| \triangleq V(f) + |f(a)|$$

defines a norm on $BV[a, b]$. To show that $\|\cdot\|$ defines a norm on $BV[a, b]$, we have

$$\|\alpha f\| = V(\alpha f) + |\alpha f(a)|$$
$$= |\alpha| V(f) + |\alpha| \, |f(a)|$$
$$= |\alpha| \, \|f\|$$

and must also show that $\|\cdot\|$ satisfies a triangle inequality

$$\|f + g\| = V(f + g) + |f(a) + g(a)|.$$

By definition,

$$V(f + g) = \sup_{P \in \mathcal{P}} \sum_{i=0}^{n-1} |f(x_{i+1}) + g(x_{i+1}) - f(x_i) - g(x_i)|$$

$$\leq \sup_{P \in \mathcal{P}} \sum_{i=0}^{n-1} \left\{ |f(x_{i+1}) - f(x_i)| + |g(x_{i+1}) - g(x_i)| \right\}$$

$$\leq \sup_{P \in \mathcal{P}} \sum_{i=0}^{n-1} |f(x_{i+1}) - f(x_i)| + \sup_{P \in \mathcal{P}} \sum_{i=0}^{n-1} |g(x_{i+1}) - g(x_i)|$$

$$= V(f) + V(g).$$

Hence,

$$\|f + g\| \leq \|f\| + \|g\|.$$

We have only left to show that

$$\|f\| = 0 \quad \Longleftrightarrow \quad f(x) = 0 \quad \text{for all} \quad x \in [a, b].$$

Clearly, if $f(x) = 0$ then $\|f\| = 0$. This follows since

$$\|f\| = \sup_{P \in \mathcal{P}} \sum_{i=0}^{n-1} |f(x_{i+1}) - f(x_i)| + |f(a)| = 0$$

*for any choice of $P \in \mathcal{P}$. On the other hand, suppose that $\|f\| = 0$ but $f(\tilde{x}) = C \neq 0$
for some $\tilde{x} \in [a, b]$. Since $\|f\| = 0$, it must be that $f(a) = 0$. Choose $P \in \mathcal{P}$ to be*

$$a = x_0 < \tilde{x} < x_2 = b.$$

That is, $\tilde{x} = x_i$. Without loss of generality we assume that $\tilde{x} \neq b$. We know that

$$\|f\| \geq \left|f(x_1) - f(x_0)\right| + \left|f(x_2) - f(x_1)\right|$$
$$= \left|f(\tilde{x}) - f(a)\right| + \left|f(b) - f(\tilde{x})\right|.$$

Consequently,

$$\|f\| \geq \left|f(x_1)\right| = \left|f(\tilde{x})\right| = \mathcal{C} \neq 0.$$

But this is a contradiction and therefore $f(x) \equiv 0$.

All of Examples 1.5.1, 1.5.2, 1.5.3, 1.5.4, and 1.5.5 provide elementary examples of complete normed vector spaces. The following examples provide increasingly complex generalizations of the Banach space $C[a, b]$ first encountered in Example 1.5.3. First, we will show that it is straightforward to extend the definition of $C[a, b]$ to accommodate more general domains.

Example 1.5.6. *In Example 1.5.3, we saw that $C[a, b]$ is a normed vector space when the norm is defined as*

$$\|f\|_{C[a,b]} = \sup_{x \in [a,b]} |f(x)| = \max_{x \in [a,b]} |f(x)|.$$

There is a clear generalization of this normed vector space that is used frequently in applications. Let Ω be an open set in $\mathbb{R}^n$. We define the set of continuous functions from Ω to $\mathbb{R}$ to be

$$C(\Omega) \stackrel{\triangle}{=} \{f : \Omega \to \mathbb{R}, \; f \text{ is continuous}\}.$$

Sometimes we will denote this set as $C^0(\Omega)$. The set $C(\Omega)$ is a normed vector space with the norm defined as

$$\|f\|_{C(\Omega)} \stackrel{\triangle}{=} \sup_{x \in \Omega} |f(x)|.$$

With this norm $C(\Omega)$ is a Banach space.

Thus, continuous functions, when endowed with an appropriate norm define not only a normed vector space, but also a Banach space. In the next example, we will see that collections of continuously differentiable functions likewise can be given a Banach space structure.

Example 1.5.7. *Let Ω be an open set in $\mathbb{R}^n$ and suppose $f : \Omega \to \mathbb{R}$. We define a multi-index m of length n to be an ordered n-tuple of integers*

$$m = \{m_1, \ldots, m_n\} \subseteq \mathbb{N}^n.$$

The length $|m|$ of the multi-index is simply the sum of the components of m,

$$|m| = \sum_{i=1}^{n} m_i.$$

Multi-indices are useful as a compact notation for defining derivatives. Let

$$D^m f \triangleq \frac{\partial^{|m|} f}{\partial x^m} \triangleq \frac{\partial^{m_1+m_2+\cdots+m_n} f}{\partial x_1^{m_1} \partial x_2^{m_2} \cdots \partial x_n^{m_n}}.$$

The set of k-times continuously differentiable functions is defined by

$$C^k(\Omega) \triangleq \{f \in C^{k-1}(\Omega) : D^m f \in C(\Omega) \quad \forall m : |m| = k\}.$$

The set of functions $C^k(\Omega)$ is a Banach space with norm

$$\|f\|_{C^k(\Omega)} \triangleq \sum_{|m| \leq k} \|D^m f\|_{C(\Omega)}.$$

Finally, we consider our most refined Banach space of classically continuously differentiable functions. These Banach spaces will be useful in characterizing important classes of domains.

Example 1.5.8. *Further refinements on the spaces of continuous functions can be made. Let $\Omega \subseteq \mathbb{R}^n$ be open and bounded and suppose $f : \Omega \to \mathbb{R}$. The function f is said to be Hölder continuous in Ω if there is a constant $c > 0$ such that*

$$|f(y) - f(x)| \leq c\,|x - y|^\alpha$$

for all $x, y \in \Omega$ and some $\alpha \in (0, 1]$. The coefficient α is the Hölder exponent. If $\alpha \equiv 1$ the function is Lipschitz continuous. A function $f \in C^m(\Omega)$ is (m, α)-Hölder continuous in Ω if there is a constant $c > 0$ and $\alpha \in (0, 1]$ such that

$$\left|D^j f(y) - D^j f(x)\right| \leq c\,|y - x|^\alpha$$

for all multi-indices $|j| \leq m$ and for all $x, y \in \Omega$. The set of (m, α)-Hölder continuous functions in Ω is a Banach space when endowed with the norm

$$\|f\|_{C^{m,\alpha}(\Omega)} \triangleq \|f\|_{C^m(\Omega)} + \max_{0 \leq |j| \leq m} \sup_{\substack{x,y \in \Omega \\ x \neq y}} \frac{\left|D^j f(y) - D^j f(x)\right|}{|y - x|^\alpha}.$$

Example 1.5.9. *Let X and Y be Banach spaces. The Banach space $C(X,Y)$ of continuous functions acting from X to Y is defined as*

$$C(X,Y) = \left\{ f : X \to Y, \quad \|f\|_{C(X,Y)} \triangleq \sup_{x \in X} \|f(x)\|_Y < \infty \right\}.$$

Similarly, the Banach space $C(\Omega_X, \Omega_Y)$ of continuous functions acting from a compact set $\Omega_X \subseteq X$ to $\Omega_Y \subseteq Y$ is defined as

$$C(\Omega_X, \Omega_Y) = \left\{ f : \Omega_X \to \Omega_Y, \quad \|f\|_{C(\Omega_X, \Omega_Y)} \triangleq \sup_{x \in \Omega_X} \|f(x)\|_Y < \infty \right\}.$$

For example, $C([a,b], \mathbb{R}^m)$ is a Banach space of continuous vector-valued functions with $\|f\|_{C([a,b], \mathbb{R}^m)} = \sup_{a \leq x \leq b} \|f(x)\|_{\mathbb{R}^m} < \infty$.

1.6   Space of Lebesgue Measurable Functions

This section discusses the functional principles and constructions employed in defining integrals on abstract spaces. These principles and constructions are widely used in various optimization applications. The reader is advised that entire monographs [8] and [5] are dedicated to this topic, and the details in the derivations are available in these texts. For at least two reasons, however, some discussion of these concepts is warranted in this book

Most modern treatments of partial differential equations in engineering and physics make use of the notion of generalized derivatives, and especially those functions whose generalized derivatives are integrable in certain classes of abstract spaces. In particular, nearly all graduate texts presenting continuum mechanics now require a working knowledge of Sobolev spaces.

The integration of Banach space valued functions is more readily understood by drawing close analogies with the derivation of integrals of real-variable functions on abstract spaces. The integration of Banach space valued functions is central to the definition of evolution equations, of both parabolic and hyperbolic type, in applications. The governing equations of an enormous class of problems in engineering and physics are expressed in terms of this class of integral.

1.6.1   Introduction to Measure Theory

With this motivation understood, we now discuss the "bare minimum" of measure and integration theory required for applications. The theory of integration in abstract spaces requires that we identify precisely which sets are "measurable." Roughly speaking, we can visualize measurable sets in $\mathbb{R}^2$, or $\mathbb{R}$, as those for which we can assign a measure of area, or of length, respectively.

Definition 1.6.1. *A collection of subsets S of a set X is a σ-algebra over X if*
1. *$X \in S$*
2. *S is closed under taking the complement of sets in S*
3. *S is closed under taking countable unions of elements of S*

Definition 1.6.2. *The Borel σ-algebra of X is the σ-algebra $\mathcal{B}$ generated by the open sets of X. An element of $\mathcal{B}$ is called a Borel set of X.*

Once we have a particular σ-algebra of subsets over a set X, we are able to define a measure space.

Definition 1.6.3. *A measure space is a triple (X, S, μ) where X is a set, S is a σ-algebra of subsets over X, and μ is a set function that satisfies two conditions:*
1. *The empty set is measurable, and has measure zero,*

$$\mu(\emptyset) = 0.$$

2. *For every sequence $\{A_i\}_{i=1}^{\infty}$ of pairwise disjoint sets contained in S, we have*

$$\mu\left(\bigcup_{i=1}^{\infty} A_i\right) = \sum_{i-1}^{\infty} \mu(A_i).$$

This last property is known as the *countable additivity* of the measure. Note that triple (X, S, μ) is called a measure space, while the pair (X, S) is referred as to *measurable space.*

Definition 1.6.4. *Let (X, S, μ) be a measure space. A set A is measurable with respect to μ if A is an element of S.*

Definition 1.6.5. *Let (X, S_X, μ_X) and (X, S_Y, μ_Y) be measure spaces. A function $f : S_X \to S_Y$ is said to be (S_X, S_Y)-measurable on X if for any set $A \subset S_Y$*

$$f^{-1}(A) \in S_X.$$

Unless stated otherwise, a real-valued function is said to be measurable if it is measurable with respect to the Borel sets.

Definition 1.6.6. *Let (X, S, μ) be a measure space. Real-valued function $f : X \to \overline{\mathbb{R}}$ is said to be μ-measurable on X if for any Borel set $A \subset \mathbb{R}$*

$$f^{-1}(A) \in S.$$

Theorem 1.6.1. *Let (X, S, μ) be a measure space. Real-valued function $f : X \to \overline{\mathbb{R}}$ is μ-measurable on X if for any $c \in \mathbb{R}$ set $\{x : f(x) > c\}$ is measurable, that is,*

$$f^{-1}((c, +\infty)) \in S.$$

Proof. The proof is obvious due to the fact that the σ-algebra, generated by the system of all intervals $(c, +\infty)$, coincides with the σ-algebra of all Borel sets on $\overline{\mathbb{R}}$.
□

The next theorem provides insight to conditions of measurability for a real-valued function defined on a closed interval $[a, b] \subset \mathbb{R}$.

Theorem 1.6.2. *A real-valued function $f : [a, b] \to \mathbb{R}$ is measurable if and only if for any $\epsilon > 0$ there exists a continuous function $\phi \in C[a, b]$ such that*

$$\mu\{x : f(x) \neq \phi(x)\} < \epsilon.$$

In the literature, there are two basic notions of convergence associated with a given measure space $(X, \mathcal{S}, \mu)$.

Definition 1.6.7. *Let $(X, \mathcal{S}, \mu)$ be a measure space. A sequence of measurable functions $\{f_n\}_{n \in \mathbb{N}}$ converges to f almost everywhere (written as a.e.) if*

$$\mu\{x : \lim_{n \to \infty} f_n(x) \neq f(x)\} = 0.$$

Definition 1.6.8. *A sequence of measurable functions $\{f_n\}_{n \in \mathbb{N}}$ converges to f with respect to (w.r.t.) measure μ if for any $\epsilon > 0$*

$$\lim_{n \to \infty} \mu\{x : |f_n(x) - f(x)| \geq \epsilon\} = 0.$$

The following two theorems establish a link between *convergence a.e.* and *convergence w.r.t. measure μ.*

Theorem 1.6.3. *Let $(X, \mathcal{S}, \mu)$ be a measure space. If a sequence of measurable functions $\{f_n\}_{n \in \mathbb{N}}$ converges to f a.e. then it converges to f w.r.t. measure μ.*

Theorem 1.6.4. *Let $(X, \mathcal{S}, \mu)$ be a measure space. If a sequence of measurable functions $\{f_n\}_{n \in \mathbb{N}}$ converges to f w.r.t. measure μ then there is a subsequence $\{f_{n_k}\}_{n_k \in \mathbb{N}} \subset \{f_n\}_{n \in \mathbb{N}}$ that converges to f a.e.*

1.6.2 Lebesgue Integral

The strategy in defining the integral of functions defined on a measure space $(X, \mathcal{S}, \mu)$ consists of two steps. First, we define the integral of finitely-valued, step functions. We subsequently define the integral of limits of finitely-valued step functions. A finitely-valued, step function is employed so frequently in this section that we refer to it as a simple function.

Definition 1.6.9. *A simple function is a finite linear combination of characteristic functions of measurable sets. That is, if f is a simple function, then it can be expressed as*

$$f = \sum_{i=1}^{N} \alpha_i \chi_{A_i}(x)$$

where the functions $\chi_{A_i}(x)$ are the characteristic functions of the measurable sets A_i, $i = 1, \ldots, N$:

$$\chi_{A_i}(x) = \begin{cases} 1 & x \in A_i, \\ 0 & x \notin A_i. \end{cases}$$

The integral of a nonnegative, simple function is defined in an intuitive fashion:

Definition 1.6.10. *Let $(X, \mathcal{S}, \mu)$ be a measure space, and let f be a nonnegative simple function having the representation*

$$f = \sum_{i=1}^{N} \alpha_i \chi_{A_i}(x)$$

where each $A_i \in \mathcal{S}$ for $i = 1, \ldots, N$. The integral of f is defined to be

$$\int_X f d\mu = \sum_{i=1}^{N} \alpha_i \mu(A_i).$$

A moments reflection shows that this definition is indeed intuitively appealing. If the measure μ corresponds to our usual notion of "area of subsets in $\mathbb{R}^2$," the integral of a nonnegative simple function is the volume under the function. In this case, it agrees with the Riemann integral defined in elementary calculus. We generalize this notion and define integrals of nonnegative functions f to be supremum over the integrals of all simple functions majorized by f. That is, we define the integral of a nonnegative function as follows:

Definition 1.6.11. *Let $(X, \mathcal{S}, \mu)$ be a measure space and let $f : X \to \overline{\mathbb{R}}_0^+$ be a nonnegative measurable function. We define the integral of f as*

$$\int_X f d\mu = \sup \left\{ \int_X \phi d\mu : \phi \text{ is a simple function, and } \phi \leq f \text{ a.e. in } X \right\}.$$

Now that, the Lebesgue integral of a nonnegative function is understood, the general definition of a Lebesgue integral is obtained by defining the absolute values of the positive and negative parts of a function f

$$|f^+|(x) = \max\{f(x), 0\}$$
$$|f^-|(x) = -\min\{f(x), 0\}.$$

Since each of the functions $|f^+|(x)$ and $|f^-|(x)$ are nonnegative, we define the Lebesgue integral of f to be

$$\int_X f(x) d\mu = \int_X |f^+|(x) d\mu - \int_X |f^-|(x) d\mu$$

whenever the integrals on the right exist (are finite). Alternatively, some authors use the following concise definition of the Lebesgue integral that does not require the separation of the positive and negative parts of a function f.

Definition 1.6.12 (Lebesgue Integral). *Let $(X, \mathcal{S}, \mu)$ be a measure space. A function $f : \mathcal{S} \to \overline{\mathbb{R}}$ is said to have Lebesgue integral on a measurable set $A \subset \mathcal{S}$ with respect to measure μ*

$$I = \int_A f(x)\, d\mu$$

if there exists a sequence of simple functions $\{f_n\}_{n \in \mathbb{N}}$ converging a.e. $x \in A$ to f such that the sequence $\{\int_A f_n(x)\, d\mu\}$ converges to I uniformly.

We will have the occasion to use the following two theorems frequently in the text. The reader is referred to [27] for the proofs.

Theorem 1.6.1 (Fatou's Lemma). *Let $(X, \mathcal{S}, \mu)$ be a measure space and suppose that f_n is a nonnegative, measurable function*

$$f_n : X \to \overline{\mathbb{R}}_0^+$$

for each $n \in \mathbb{N}$. If

$$f_n \to f \quad a.e. \ in \ X$$

then we have

$$\int_X f d\mu \leq \liminf_{n \to \infty} \int_X f_n d\mu$$

It should be noted that the conclusion of Fatou's Lemma may also be written as

$$\int_X (\liminf_{n \to \infty} f_n) d\mu \leq \liminf_{n \to \infty} \int_X f_n d\mu$$

which may be easier to remember.

Theorem 1.6.2 (Lebesgue's Dominated Convergence Theorem). *Let $(X, \mathcal{S}, \mu)$ be a measure space, and suppose that $\{f_n\}_{n=1}^{\infty}$ is a sequence of measurable functions*

$$f_n : X \to \overline{\mathbb{R}}$$

such that

$$|f_n(x)| < g(x) \quad \forall x \in X$$

for some integrable function g. If

$$f_n \to f \quad a.e. \ in \ X$$

then

$$\int_X f_n d\mu \to \int_X f d\mu.$$

1.6.3 Measurable Functions

This section presents two theorems on measurability conditions for various functions, which are often encountered in mathematical modelling of various applications in mechanics and optimal control.

Theorem 1.6.5. *Let $(T, \mathcal{S}, \mu)$ be a measure space, X be a separable topological space and let $f : T \times X \to \mathbb{R}$ be such that for each $x \in X$*

$$f(\cdot, x) : T \to \mathbb{R} \quad is \quad \mu\text{-measurable}$$

and for each $t \in T$

$$f(t, \cdot) : X \to \mathbb{R}$$

is continuous and bounded. Then the function

$$g : T \to \mathbb{R}$$

$$g(t) \triangleq \sup_{x \in X} f(t, x)$$

is μ-measurable.

Proof. Since X is a separable topological space, there is a countable set $\{x_1, x_2, \ldots\}$ that is dense in X. We claim that

$$g(t) \triangleq \sup_{x \in X} f(t, x) = \sup_{i \in \mathbb{N}} f(t, x_i).$$

It is clear that

$$s \triangleq \sup_{x \in X} f(t, x) \geq \sup_{i \in \mathbb{N}} f(t, x_i)$$

simply because the supremum over a set must be greater than the supremum over any of its subsets. By the definition of the supremum, there must be a sequence $\{\xi_k\}_{k \in \mathbb{N}} \subseteq X$ such that

$$f(t, \xi_k) \to s$$

as $k \to \infty$. Since $\{x_i\}_{i \in \mathbb{N}}$ is dense in X, for each k there is a sequence $\{x_{k_j}\}_{j \in \mathbb{N}} \subseteq \{x_i\}_{i \in \mathbb{N}}$ such that

$$x_{k_j} \to \xi_k \quad as \quad j \to \infty.$$

We can write

$$\left| s - f(t, x_{k_j}) \right| \leq \left| s - f(t, \xi_k) \right| + \left| f(t, \xi_k) - f(t, x_{k_j}) \right|.$$

By the continuity of $f(t, \cdot)$ on X, we have

$$f(t, x_{k_j}) \to f(t, \xi_k)$$

as $k_j \to \infty$. Thus,

$$g(t) = \sup_{x \in X} f(t, x) = \sup_{i \in \mathbb{N}} f(t, x_i).$$

For any given constant $c \in \mathbb{R}$,

$$g(t) \leq c \quad \text{if and only if} \quad f(t, x_i) \leq c \quad \text{for all} \quad i \in \mathbb{N}.$$

We can write

$$g^{-1}((-\infty, c]) = \{t \in T : g(t) \leq c\}$$
$$= \bigcap_{i \in \mathbb{N}} \{t \in T : f(t, x_i) \leq c\}.$$

By hypothesis, each set on the right in μ-measurable. The set $g^{-1}((-\infty, c]) \in \mathcal{S}$ for each $c \in \mathbb{R}$ and we conclude that g is measurable. $\qquad \square$

Theorem 1.6.6. *Let (X, d) be a separable metric space, $(T, \mathcal{S}, \mu)$ a measure space and $f : T \to X$. Then the following are equivalent*

- *f is μ-measurable*
- *$g \circ f$ is μ-measurable for every continuous $g : X \to \mathbb{R}$*
- *The function $t \to d(f(t), x)$ is μ-measurable $\forall\, x \in X$.*

Proof. See [33, p. 71]. $\qquad \square$

1.7   Hilbert Spaces

As a final example of a general class of an abstract space, we consider inner product spaces. This class of spaces has more mathematical structure than any of the previously discussed spaces. In summary, every inner product space is

- a topological space,
- a metric space, and
- a normed vector space.

In addition, an inner product space defines an attribute of its elements that does not arise in the definitions of general topological, metric or normed vector spaces. The notion of "orthogonality" of the elements is defined in an inner product space.

Definition 1.7.1. *Let X be a vector space. An inner product is a mapping $(\cdot, \cdot) : X \times X \to \mathbb{C}$. The inner product must satisfy*

(IP1)	$(x + y, z) = (x, z) + (y, z)$	$\forall\, x,\, y,\, z \in X$	
(IP2)	$(\alpha x, y) = \alpha(x, y)$	$\forall\, x,\, y \in X$	$\forall\, \alpha \in \mathbb{C}$
(IP3)	$(x, y) = \overline{(y, x)}$	$\forall\, x,\, y \in X$	
(IP4)	$(x, x) \geq 0$	$\forall\, x \in X$	
	and $(x, x) = 0 \iff$	$x = 0.$	

Several obvious properties of an inner product can easily be deduced from these axioms

$$(x, \alpha\, y) = \overline{\alpha}(x, y)$$
$$(x, y + z) = (x, y) + (x, z).$$

A vector space equipped with an inner product is an inner product space. An inner product space has as much structure as a normed vector space, and hence a topological and metric space, because we can always define a norm from the inner product.

Definition 1.7.2. *The inner product generates a norm on X given by $\|x\| = \sqrt{(x,x)}$. Hence, all inner product spaces are normed vector spaces, but the converse is not generally true.*

Among the more common attributes of a Hilbert space we will frequently use the Cauchy-Schwarz Inequality, and the continuity of the inner product.

Theorem 1.7.1 (Cauchy-Schwarz Inequality). *Let X be an inner product space and suppose that $x, y \in X$. For any x, $y \in X$, we have*

$$|(x,y)| \leq \|x\| \, \|y\|.$$

Theorem 1.7.2. *The inner product is continuous. That is, if $x_n \to x \in H$ and $y_n \to y \in H$, the inner product $(x_n, y_n) \to (x,y) \in \mathbb{R}$ or $\mathbb{C}$.*

And, as in the case of a general normed vector space, the metric induced via the inner product may, or may not, induce a complete metric space structure. That is, an inner product space is complete if the metric induced from its inner product is complete over the set H.

Definition 1.7.3. *A Hilbert Space is a complete inner product space.*

The introduction of the inner product allows us to introduce the concept of orthogonality of elements in an inner product space. The following theorem shows that this idea can lead to very intuitive constructions in inner product spaces. We define the orthogonal complement $A^\perp$ of a set $A \subseteq H$ to be

$$A^\perp = \left\{ x \in H : (x,a) = 0 \quad \forall\, a \in A \right\}.$$

The orthogonal complement appears in many important theorems that are stated within the Hilbert space framework. For example, a set is dense in a Hilbert space if and only if the only element that is orthogonal to the dense set is the zero vector. The concepts of *linear subspace* along with *orthogonal complement* induce an additional structure in a Hilbert space.

Definition 1.7.4. *A set $A \subseteq H$ is a linear subspace of a Hilbert space H if $\alpha_1 x_1 + \alpha_2 x_2 \in A$ for any x_1, $x_2 \in A$ and α_1, $\alpha_2 \in \mathbb{R}$.*

The following two propositions establish a connection between notions of *linear subspace* and *orthogonal complement*

Proposition 1.7.1. *If A is an arbitrary set of a Hilbert space H then its orthogonal complement $A^\perp$ is a closed linear subspace of H.*

Proof. $A^\perp$ is a linear subspace if for any x_1, $x_2 \in A^\perp$ and $\alpha_1, \alpha_2 \in \mathbb{R}$ we show that $\alpha_1 x_1 + \alpha_2 x_2 \in A^\perp$. Indeed, based on inner product axioms (IP1) and (IP2), for any $a \in A$ we have

$$(\alpha_1 x_1 + \alpha_2 x_2, a) = \alpha_1(x_1, a) + \alpha_2(x_2, a) = 0.$$

To show that $A^\perp$ is closed, consider a sequence $\{x_n\}_{n \in \mathbb{N}} \in A^\perp$ such that $x_n \to x$ and $(x_n, a) = 0$ for all $a \in A$ and $n \in \mathbb{N}$. According to Theorem 1.7.2, the inner product is continuous, and consequently, we obtain

$$\lim_{n \to \infty} (x_n, a) = (x, a) = 0.$$

But this means that $x \in A^\perp$, and thus, closedness of $A^\perp$ is shown.   □

The next theorem plays an important role in the theory of Hilbert spaces.

Theorem 1.7.3. *Let $A \subseteq H$ be a closed linear subspace of a Hilbert space H and $A^\perp$ be its orthogonal complement. Then any element $x \in H$ can uniquely be represented as $x = a + a'$, where $a \in A$ and $a' \in A^\perp$. We write $H = A \oplus A^\perp$.*

Proof. First we establish the existence of the representation $x = a + a'$ for any $x \in H$, where $a \in A$ and $a' \in A^\perp$. Let $d = \rho(x, A)$, and let the sequence $\{a_n\}_{n \in \mathbb{N}} \in A$ be defined such that $||x - a_n|| < d^2 + \frac{1}{n^2}$. This sequence is convergent. Indeed,

$$||a_n - a_m||^2 = 2\left(||x - a_n||^2 + ||x - a_m||^2\right) - ||(x - a_n) + (x - a_m)||^2$$

$$= 2\left(||x - a_n||^2 + ||x - a_m||^2\right) - 4\left|\left|x - \frac{a_n + a_m}{2}\right|\right|^2$$

$$< 2\left(d^2 + \frac{1}{n^2} + d^2 + \frac{1}{m^2}\right) - 4d^2$$

$$= \frac{1}{n^2} + \frac{1}{m^2}.$$

Since H is a complete space, there exists a such that $\lim_{n \to \infty} a_n = a$. Moreover, since A is closed, we have $a \in A$. Based on the continuity of the inner product, we also have $\lim_{n \to \infty} ||x - a_n|| = ||x - a|| = d$. Now let us show that $a' = x - a \in A^\perp$. Let $b \in A$. Since A is a linear subspace, $a + \lambda b \in A$ for any $\lambda \in \mathbb{R}$. By definition of d, we obtain a chain of equivalent inequalities

$$||x - (a + \lambda b)||^2 \geq d^2$$

$$||(x - a) - \lambda b||^2 \geq ||x - a||^2$$

$$||a' - \lambda b||^2 \geq ||a'||^2$$

$$(a', a') - 2\lambda(a', b) + \lambda^2(b, b) \geq (a', a')$$

$$\lambda^2(b, b) \geq 2\lambda(a', b)$$

which should hold for all $\lambda \in \mathbb{R}$. Substituting $\lambda = \frac{(a',b)}{(b,b)}$ into the last inequality, we obtain $\frac{|(a',b)|^2}{(b,b)} \geq 2\frac{|(a',b)|^2}{(b,b)}$. But it follows that $|(a',b)|^2 \leq 0$, which can be true only if $(a',b) = 0$. Since b was chosen arbitrarily, $(a',b) = 0$ for any $b \in A$. Thus we proved that $a' \in A^\perp$ and the existence of $x = a + a'$, where $a \in A$ and $a' \in A^\perp$.

To prove uniqueness of this representation, suppose that x can also be represented as $x = b + b'$, where $b \in A$ and $b' \in A^\perp$. Consequently, $x = a + a' = b + b'$, which can be reduced to $a - b = b' - a'$. Because A and $A^\perp$ are linear subspaces, we have $a - b \in A$ and $b' - a' \in A^\perp$, and hence $(a - b, b' - a') = 0$. However, since $a - b = b' - a'$, we obtain $(a - b, a - b) = 0$ and $(b' - a', b' - a') = 0$. It follows that $a = b$ and $a' = b'$, and the theorem is proved. $\qquad\square$

Proposition 1.7.2. *If A is a linear subspace of a Hilbert space H then $(A^\perp)^\perp = \overline{A}$.*

Proof. Let $B = A^\perp$. By definition of the subspace B, any element $a \in A$ is orthogonal to B, i.e., $a \perp B$. This means that $a \in B^\perp$, and consequently $A \subseteq B^\perp$. According to Proposition 1.7.1, the space $B^\perp$ is closed. Hence, $\overline{A} \subseteq B^\perp$. On the other hand, $(\overline{A})^\perp \subseteq B$. Indeed, suppose $a' \in (\overline{A})^\perp$, i.e. $a' \perp \overline{A}$. Consequently, $a' \perp A \subseteq \overline{A}$. But this means that $a' \in A^\perp$, and thus $(\overline{A})^\perp \subseteq A^\perp = B$. Now consider an arbitrary element $x \in B^\perp$. By Theorem 1.7.3, the Hilbert space H can be represented as $H = \overline{A} \oplus (\overline{A})^\perp$ and the element x is represented as $x = a + a'$, where $a \in \overline{A}$ and $a' \in (\overline{A})^\perp$. Since $a \in \overline{A} \subseteq B^\perp$, $x \in B^\perp$ and $B^\perp$ is a linear subspace, we have $a' = x - a \in B^\perp$. However $a' \in (\overline{A})^\perp \subseteq B$. But $a' \in B$ and $a' \in B^\perp$ can hold only for $a' = 0$. Consequently, $x = a \in \overline{A}$ and $B^\perp \subseteq \overline{A}$. From $\overline{A} \subseteq B^\perp$ and $B^\perp \subseteq \overline{A}$ we conclude that $\overline{A} = B^\perp = (A^\perp)^\perp$, and the proposition is proved. $\qquad\square$

Theorem 1.7.4. *Let H be a Hilbert space. A set V is dense in H if and only if $V^\perp = \{0\}$.*

Proof. We will show that if V is dense in H, then $V^\perp = \{0\}$. Suppose V is dense in H and let $v_0 \in V^\perp$. Since $v_0 \in H$ there is a sequence $\{v_k\}_{k\in\mathbb{N}} \subseteq V$ such that

$$v_k \to v_0 \quad \text{in} \quad H.$$

Using the Cauchy-Schwarz inequality and the fact that $(v_k, v_0) = 0$, we can directly compute

$$\begin{aligned} \|v_0\|^2 &= (v_0, v_0) - (v_k, v_0) \\ &= (v_0 - v_k, v_0) \\ &\leq \|v_0 - v_k\| \, \|v_0\|. \end{aligned}$$

Consequently, we have

$$\|v_0\| \leq \|v_0 - v_k\| \to 0 \quad \text{as} \quad k \to \infty.$$

Hence, $v_0 \equiv 0$ and $V^\perp = \{0\}$. $\qquad\square$

To show that if $V^\perp = \{0\}$ then V is dense is left as an exercise.

This theorem has immediate applicability. It is used in the next theorem to show that every Hilbert Space has a dense orthonormal set.

Proposition 1.7.3. *Let H be a nontrivial Hilbert space. Then H possesses a dense orthonormal collection of vectors.*

Proof. In this case we construct a partially ordered set by taking Y to be the subsets of H comprised of mutually orthonormal elements. Specifically,

$$\mathcal{Y} \in Y \quad \Longleftrightarrow \quad \mathcal{Y} = \{y_\gamma \in H : \gamma \in \Gamma, \quad y_{\gamma_1} \perp y_{\gamma_2} \quad \forall\, \gamma_1, \gamma_2 \in \Gamma\}$$

for some (perhaps uncountable) index set Γ. If H is nontrivial, then $\mathcal{Y}$ is nonempty since

$$x \neq 0, \qquad x \in H \quad \Longrightarrow \quad \left\{ \frac{x}{\|x\|} \right\} \in Y.$$

Again, set inclusion defines a partial ordering on Y

$$\mathcal{Y} \leq \mathcal{Z} \quad \Longleftrightarrow \quad \mathcal{Y} \subseteq \mathcal{Z}.$$

Moreover, if $\{\mathcal{Y}_\alpha\}_{\alpha \in A}$ is a totally ordered set, then the set

$$\mathcal{Y}_u = \bigcup_{\alpha \in A} \mathcal{Y}_\alpha$$

is an upper bound for the chain. By Zorn's lemma, there is a maximal element $\mathcal{Y}_m \in Y$. We now claim that $\mathcal{Y}_m$ is dense in H. Recall that a set F is dense in the Hilbert space H if and only if

$$F^\perp = \{0\}.$$

Suppose that $\exists\, y_0 \in \mathcal{Y}_m^\perp$ and $y_0 \neq 0$. Then the set $\mathcal{Y}_m \cup \{y_0\}$ is related to $\mathcal{Y}_m$

$$\mathcal{Y}_m \subseteq \mathcal{Y}_m \cup \{y_0\}.$$

But this contradicts the maximality of $\mathcal{Y}_m$ and the proposition is proved.   □

The last theorem becomes more constructive for *separable* Hilbert spaces.

Definition 1.7.5. *A Hilbert space H is separable if there is a countable, everywhere dense set in H.*

Theorem 1.7.5. *Let H be a separable Hilbert space. Then H possesses a countable orthonormal collection of vectors.*

In the case of infinite-dimensional separable Hilbert spaces, a countable orthonormal collection is also complete.

Theorem 1.7.6. *Let H be an infinite-dimensional separable Hilbert space. Then H possesses a complete countable orthonormal collection of vectors.*

The most remarkable property of separable Hilbert spaces is that all separable Hilbert spaces are isomorphic.

Theorem 1.7.7. *Any two separable Hilbert spaces are isomorphic to each other.*

Chapter 2

Linear Functionals and Linear Operators

The introduction of the algebraic operations that define a vector space enable us to introduce an important class of functions on normed linear spaces: the class of linear operators. We denote the domain $\text{dom}(T)$ of the operator $T : X \to Y$ to be the set of $x \in X$ such that $T(x) \in Y$. That is,

$$\text{dom}(T) = \{x \in X : T(x) \in Y\}.$$

Likewise, the range of the operator $T : X \to Y$ is

$$\text{range}(T) = \{y \in Y : \exists\, x \in X \text{ such that } Tx = y\}.$$

Definition 2.0.6. *An operator T mapping elements of the vector space X to the vector space Y is a linear operator if $\forall\, x, y \in \text{dom}(T)$ and scalars α*

$$
\begin{aligned}
T(x + y) &= Tx + Ty \\
T(\alpha x) &= \alpha T x.
\end{aligned}
$$

It is an elementary exercise to show that the set of all linear operators that act between two vector spaces X and Y form a vector space themselves. This vector space is denoted by $L(X, Y)$, where the operations of "vector addition" (addition of linear operators) and scalar multiplication are defined in the obvious way:

$$
\begin{aligned}
(T_1 + T_2)x &\equiv T_1(x) + T_2(x) \\
(\alpha T_1)x &\equiv T_1(\alpha x).
\end{aligned}
$$

Definition 2.0.7. *Let X and Y be vector spaces with dimension n_x and n_y, respectively. Then the set of all linear mappings from X into Y, denoted $L(X, Y)$, is a vector space of dimension $n_x \times n_y$.*

Not only can we define a new vector space $L(X, Y)$ from the two vector spaces X and Y, we can endow a subset $\mathcal{L}(X, Y) \subseteq L(X, Y)$ with a norm; the *operator norm*. We define the operator norm by first defining the important class of linear operators that map bounded sets of vectors in the domain into bounded sets of vectors in the range.

Definition 2.0.8. *Let X and Y be normed vector spaces and $T \colon \operatorname{dom}(T) \subset X \to Y$ a linear operator. We say that T is a bounded linear operator if there is a real number c such that*

$$\|Tx\| \le c \, \|x\|, \quad \forall x \in \operatorname{dom}(T) \,.$$

For each *bounded*, linear operator we define a norm.

Definition 2.0.9. *The operator norm of the bounded linear operator T is defined by*

$$\|T\| = \sup_{x \ne 0} \frac{\|Tx\|}{\|x\|} \quad \forall x \in \operatorname{dom}(T) \,.$$

The set of all bounded linear operators from X into Y is denoted by $\mathcal{L}(X, Y)$. To emphasize a point, recall that the following definition characterizes continuity of a mapping T acting between normed vector spaces.

Definition 2.0.10. *Let $T \colon \operatorname{dom}(T) \subseteq X \longrightarrow Y$ be an operator, not necessarily linear, where $\operatorname{dom}(T) \subset X$ and X and Y are normed vector spaces. Then T is continuous at a point $x_0 \in \operatorname{dom}(T)$ if for every $\epsilon > 0$ there is a $\delta > 0$ such that*

$$\|x - x_0\| < \delta \quad \Longrightarrow \quad \|Tx - Tx_0\| < \epsilon \,.$$

If T is continuous at every $x \in \operatorname{dom}(T)$, then T is a continuous operator.

Because normed vector spaces are indeed examples of topological spaces, this definition is but one special case of our general definition. Interestingly enough, for linear operators acting between normed vector spaces, the notion of continuity coincides with that of boundedness.

Theorem 2.0.8. *Let $T \colon \operatorname{dom}(T) \subseteq X \to Y$ be a linear operator, where X, Y are normed spaces. Then T is continuous if and only if T is bounded.*

In our definition of a linear operator, we have allowed that the operator need not be defined on the entire underlying vector space. That is, $\operatorname{dom}(T) \subseteq X$, and we have not insisted that T exhausts X. If we are given an operator T defined on some subspace of X, it is often necessary to define an extension of the operator T.

Definition 2.0.11. *Let $T \colon \operatorname{dom}(T) \subseteq X \to Y$, where X and Y are normed vector spaces. The extension of T to all of X is an operator*

$$\tilde{T} \colon X \to Y, \quad \text{such that} \quad \tilde{T}|_{\operatorname{dom}(T)} = T \,.$$

The notation $\tilde{T}|_{\operatorname{dom}(T)}$ denotes the restriction of $\tilde{T}$ to $\operatorname{dom}(T)$.

The following theorem embodies a fundamental tool from mathematical analysis that we will have frequent occasion to use in the chapters that follow. This theorem states that a bounded, linear operator defined on a dense subset of a normed vector space has a "canonical" extension.

Theorem 2.0.9. *Consider the bounded linear operator* $T \colon \mathrm{dom}(T) \subseteq X \to Y$, *where* Y *is a Banach space and* X *is a normed vector space. Then* T *has a bounded linear extension* $\tilde{T}$ *such that*

$$\tilde{T} : \overline{\mathrm{dom}(T)} \subseteq X \to Y$$

$$\|\tilde{T}\| = \|T\|.$$

Note that if $\mathrm{dom}(T)$ *is dense in* X *the operator is extended to all of* X.

2.1 Fundamental Theorems of Analysis

If X is a linear vector space, the set of all operators that map X into $\mathbb{R}$ are denoted the *functionals* on the space X.

Definition 2.1.1. *A linear functional* f *on a vector space* X *is an element of* $\mathcal{L}(X, \mathbb{R})$.

2.1.1 Hahn-Banach Theorem

Of all the foundational theorems of functional analysis, the Hahn-Banach theorem can be the most mystifying to beginning students. In principle, the Hahn-Banach theorem is simple to describe: it provides a means of constructing extensions to linear functionals defined on subspaces of Banach spaces. Apparently, many students would question why we want to do this in the first place. In many applications, this is a common requirement and crucial to the solution of many practical problems in science, engineering and control theory. There are many forms of the Hahn-Banach theorem, and we first summarize what may be the most general form considered in this book. Its proof relies on Zorn's lemma.

Theorem 2.1.1 (Hahn-Banach Theorem). *Let* X *be a real vector space and let* $f : X \to \mathbb{R}$ *be positively homogeneous,*

$$f(\alpha x) = \alpha f(x)$$

for all $\alpha \geq 0$ *and all* $x \in X$, *and subadditive,*

$$f(x + y) \leq f(x) + f(y)$$

for all $x, y \in X$. *Let* L_0 *be a linear functional defined on a subspace* $Y \subseteq X$ *that is majorized by* f *over* Y:

$$L_0(x) \leq f(x) \qquad \forall \ x \in Y.$$

Then there is a linear extension L *of* L_0

$$L : X \to \mathbb{R}$$
$$L(x) = L_0(x) \qquad \forall \ x \in Y$$

and L is majorized by f over X

$$L(x) \leq f(x) \qquad \forall\, x \in X.$$

Proof. Let S be the set of all linear extensions of L_0 that are majorized by f. That is,

$$S = \Big\{ g \in L\big(\mathrm{dom}(g), \mathbb{R}\big) : \mathrm{dom}(L_0) \subseteq \mathrm{dom}(g),\ L_0(x) = g(x) \quad \forall\, x \in \mathrm{dom}(L_0),$$

$$g(x) \leq f(x) \quad \forall\, x \in \mathrm{dom}(g) \Big\}.$$

S is a nonempty set since $L_0 \in S$. Moreover, we can define a partial ordering $\gtrsim$ on the set S. If $a, b \in S$, then we say

$$a \gtrsim b$$

if

$$\mathrm{dom}(a) \supseteq \mathrm{dom}(b)$$
$$a(x) = b(x) \qquad \forall\, x \in \mathrm{dom}(b).$$

It is straightforward to show that $(S, \gtrsim)$ is a partially ordered set. We want to apply Zorn's lemma to conclude that the set S contains a maximal element. Suppose $\mathcal{C}$ is any totally ordered subset in S. For the chain $\mathcal{C}$, define the function l_c such that

$$\mathrm{dom}(l_c) \in \bigcup_{g \in \mathcal{C}} \mathrm{dom}(g)$$

and

$$l_c(x) = g(x) \qquad x \in \mathrm{dom}\,(g). \tag{2.1}$$

Clearly, the definition in Equation (2.1) makes sense since $x \in \mathrm{dom}(l_c)$ implies that there is some $g \in \mathcal{C}$ with $x \in \mathrm{dom}(g)$. If there are $g_1, g_2 \in \mathcal{C}$, $x \in \mathrm{dom}(l_c)$, and

$$x \in \mathrm{dom}\,g_1 \cap \mathrm{dom}\,g_2$$

the definition in (2.1) remains clear. Since $\mathcal{C}$ is a chain either $g_1 \gtrsim g_2$ or $g_2 \gtrsim g_1$. Thus,

$$l_c(x) = g_1(x) = g_2(x).$$

We claim that l_c is an upper bound for $\mathcal{C}$. That is, we must show that

$$l_c \gtrsim g \qquad \forall\, g \in \mathcal{C}.$$

But this is certainly true since for any $g \in \mathcal{C}$, we have

$$\mathrm{dom}(l_c) = \bigcup_{h \in \mathcal{C}} \mathrm{dom}(h) \supseteq \mathrm{dom}(g)$$

and

$$l_c(x) = g(x) \qquad \text{for } x \in \mathrm{dom}(g).$$

Consequently we have shown that any arbitrary chain $\mathcal{C}$ in S has an upper bound. By Zorn's lemma S has a maximal element. Denote the maximal element of S as L. By the definition S, the element L is linear, it is an extension of L_0 and it is majorized by f. That is

$$L(x) = L_0(x) \qquad \forall\, x \in Y$$
$$L(x) \le f(x) \qquad \forall\, x \in \mathrm{dom}(L) \subseteq X.$$

If we can show that $\mathrm{dom}(L) = X$, the proof will be complete. As conventional in constructions via Zorn's lemma, we will argue by contradiction. Suppose that there is some $x_0 \in X$ such that $x_0 \notin \mathrm{dom}(L)$. Since $\mathrm{dom}(L)$ is a vector space, we construct a linear functional on the vector space Z

$$Z \triangleq \mathrm{dom}(L) \oplus \mathrm{span}\{x_0\}.$$

Fix a constant $a \in \mathbb{R}$. For any $x \in \mathrm{dom}(L)$, define the functional $\tilde{L}$ such that

$$\tilde{L}(x + cx_0) = L(x) + ac.$$

The functional $\tilde{L}$ is linear on Z. Any two elements $z_1, z_2 \in Z$ have the unique representation

$$z_1 = x_1 + c_1 x_0$$
$$z_2 = x_2 + c_2 x_0$$

for $x_1, x_2 \in \mathrm{dom}(L)$ and $c_1, c_2 \in \mathbb{R}$. By definition, we have

$$\begin{aligned}
\tilde{L}(z_1 + \gamma z_2) &= L(x_1 + \gamma x_2) + a(c_1 + \gamma c_2) \\
&= L(x_1) + ac_1 + \gamma\big(L(x_2) + ac_2\big) \\
&= \tilde{L}(z_1) + \gamma \tilde{L}(z_2).
\end{aligned}$$

Moreover, $\tilde{L}$ is an extension of L since

$$\mathrm{dom}(L) \subseteq \mathrm{dom}(\tilde{L}) = z.$$

In fact, we would have

$$\tilde{L} \in S \quad \text{and} \quad \tilde{L} \gtrsim L$$

if we could show that $\tilde{L}$ is majorized by f

$$\tilde{L}(x) \le f(x) \qquad \forall\, x \in \mathrm{dom}(\tilde{L}).$$

Consider any x and y in $\mathrm{dom}\,(L)$. We can compute

$$\begin{aligned}
L(x) - L(y) &= L(x - y) \\
&\le f(x - y) \\
&\le f(x + x_0 - x_0 - y) \\
&\le f(x + x_0) + f(-x_0 - y).
\end{aligned}$$

In this sequence of equations, we have used the linearity of L, the majorization of L by f, and the subadditivity of f. We can rearrange the above inequality to make

$$-L(y) - f(-x_0 - y) \le -L(x) + f(x + x_0).$$

We note that this equation holds for all $x \in$ dom (L) and for all $y \in$ dom (L). Recall that our definition of the functional $\tilde{L}$ is

$$\tilde{L}(x + cx_0) = L(x) + ac$$

but we have not yet selected the constant $a \in \mathbb{R}$. We always have

$$\sup_{y \in \text{dom }(L)} \left(-L(y) - f(-x_0 - y) \right) \le \inf_{x \in \text{dom}(L)} \left(-L(x) + f(x + x_0) \right).$$

Choose the constant a such that

$$\sup_{y \in \text{dom }(L)} \left(-L(y) - f(-x_0 - y) \right) \le a \le \inf_{x \in \text{dom }(L)} \left(-L(x) + f(x + x_0) \right). \quad (2.2)$$

Now the value of the functional $\tilde{L}$ on an element

$$x + cx_0 \in Z \quad (2.3)$$

can be studied for the case when $c > 0$, $c = 0$, and $c < 0$. Suppose $c = 0$. Then $x \in$ dom (L) and

$$\tilde{L}(x) = L(x) \le f(x)$$

and $\tilde{L}$ is majorized by f in this case. On the other hand suppose $c > 0$. From Equation (2.2). We know that

$$a \le -L(\xi) + f(\xi + x_0)$$

for any choice of $\xi \in$ dom (L). Choose $\xi = \frac{1}{c}x$ where c and x are given in Equation (2.3). We know that

$$a \le -L\left(\frac{1}{c}x\right) + f\left(\frac{1}{c}x + x_0\right).$$

If we multiply by the positive number c, and rearrange the terms, we obtain

$$L(x) + ca \le f(x + cx_0)$$

or

$$\tilde{L}(z) \le f(z) \qquad \forall\, z \in \text{dom }(\tilde{L}).$$

But this means $\tilde{L} \in S$ and

$$\tilde{L} \gtrsim L$$

which is a contradiction to the maximality of L. Now suppose $c < 0$. From Equation (2.2), we can write

$$-L(\xi) - f(-x_0 - \xi) \le a \quad (2.4)$$

for any choice of $\xi \in \text{dom}\,(L)$. Again, with the choice of

$$z = x + cx_0.$$

Choose $\xi = \frac{1}{c}x$ in Equation (2.4):

$$-L\left(\frac{1}{c}x\right) - f\left(-x_0 - \frac{1}{c}x\right) \le a.$$

If we multiply the above equation by the positive number $(-c)$ we get

$$L(x) - f(x + cx_0) \le -ca$$
$$L(x) + ca \le f(x + cx_0)$$
$$\tilde{L}(z) \le f(z).$$

Again, we get $\tilde{L} \in S$ and $\tilde{L} \ge L$, which is a contradiction of the maximality of L. Hence, our assumption that $\text{dom}\,(L) \ne X$ is false. We have

$$\text{dom}\,(L) = X$$

and the theorem is proved. $\square$

2.1.2 Uniform Boundedness Theorem

In terms of application, the Uniform Boundedness Theorem may find more direct application for engineers and scientists than the Baire category theorem in Section 1.3.4 or the Hahn-Banach theorem in Section 2.1. Its proof, at least the form presented here, employs the Baire Category Theorem. Philosophically, the Uniform Boundedness Theorem enables us to utilize "pointwise bounds" on families of bounded linear operators to conclude that we actually have "uniform bounds" on the same family.

Theorem 2.1.2 (Uniform Boundedness Theorem). *Let $\{L_k\}_{k\in\mathbb{N}}$ be a sequence of bounded linear operators on the Banach space X. Suppose that for each $x \in X$, there exists a constant c_x (depending on x) such that*

$$\|L_k x\| \le c_x \qquad \forall\, k \in \mathbb{N}. \tag{2.5}$$

Then the family $\{L_k\}_{k\in\mathbb{N}}$ is uniformly bounded in the sense that there is a fixed constant C such that

$$\|L_k\| \le C \qquad \forall\, k \in \mathbb{N}.$$

Proof. For each integer $k \in \mathbb{N}$, define the set C_k

$$C_k = \{x \in X : \quad \|L_n x\| \le k \qquad \forall\, n \in \mathbb{N}\}.$$

Each C_k is closed. Perhaps the simplest way to see this is to note that the map f

$$f : x \mapsto \|L_n x\|$$

is a continuous map from X to $\mathbb{R}$. Suppose $\{x_j\}_{j \in \mathbb{N}} \subseteq C_k$ and $x_j \to x$. Then

$$\|L_n x_j\| \to \|L_n x\| \le k$$

and $x \in C_k$. In addition

$$X = \bigcup_{k=1}^{\infty} C_k$$

by the pointwise bound in Equation (2.5). By the Baire category theorem, there must be some $k_0 \in \mathbb{N}$ such that C_{k_0} contains an open ball

$$B(x_0, r_0) \subseteq C_{k_0}.$$

If x is any element of X, we can define

$$y = x_0 + \frac{1}{2} r_0 \frac{x}{\|x\|}.$$

Note that

$$\|y - x_0\| \le \frac{1}{2} r_0 \quad \Longrightarrow \quad y \in B(x_0, r_0) \subseteq C_{k_0}$$

and that $x_0, y \in C_{k_0}$ implies that

$$\|L_n x_0\| \le k_0$$

$$\|L_n y\| \le k_0$$

for all $n \in \mathbb{N}$. By construction, we have

$$x = 2 \frac{\|x\|}{r_0} (y - x_0)$$

and

$$\begin{aligned}
\|L_n x\| &= 2 \frac{\|x\|}{r_0} \|L_n(y - x_0)\| \\
&\le 2 \frac{\|x\|}{r_0} (\|L_n y\| + \|L_n x_0\|) \\
&\le \frac{4k_0}{r_0} \|x\|.
\end{aligned}$$

This implies that $4k_0/r_0$ is an upper bound for $\|L_n\|$ and the theorem is proved. $\square$

2.1.3 The Open Mapping Theorem

We studied several alternative characterizations of continuity of functions in the previous chapter. Perhaps the most abstract definition is cast in terms of inverse images of open sets. A function $f : X \to Y$ is continuous according to this definition if and only if for every open set $\Theta \subseteq Y$ the inverse image $f^{-1}(\Theta) \subseteq X$ is an open set. The reader should carefully note that there is no guarantee that a continuous function will map open sets in X into open sets in Y. Functions having this particular property are referred to as open mappings.

Definition 2.1.2. *Let* $f : \operatorname{dom}(f) \subseteq X \to Y$ *where* X *and* Y *are topological spaces. The function* f *is called an open mapping if the image under* f *of every open set in the domain of* f *is an open set in* Y.

One of the most important applications of open mappings is that they can sometimes provide a convenient method for establishing the boundedness of certain invertible operators. This fact is summarized in the celebrated *Open Mapping Theorem*, which is also known as *Banach-Schauder Theorem*.

Theorem 2.1.3 (Open Mapping Theorem). *Let* L *be a bounded linear operator from a Banach space* X *onto a Banach space* Y. *Then* L *is an open mapping.*

This theorem is extremely powerful. In fact, it is an immediate consequence of the theorem that if the bounded linear operator is invertible, the inverse is necessarily continuous. Moreover, since the inverse of a linear operator is itself linear, the Open Mapping Theorem guarantees that if the inverse exists, it is bounded.

The proof of the Open Mapping Theorem is lengthy. We will first derive an intermediate result that will facilitate the proof of the theorem. Similar approaches can be found in [3, 15, 30].

Proposition 2.1.1. *Suppose* L *is a bounded linear operator mapping the Banach space* X *onto the Banach space* Y. *Let* $B_1(0)$ *be the open ball of radius* 1 *centered at* $0 \in X$

$$B_1(0) = \{x \in X : \|x\| < 1\}.$$

The image $L\big(B_1(0)\big)$ *contains an open ball* θ *containing the origin in* Y

$$0 \in \theta \subseteq L\big(B_1(0)\big).$$

Proof. The proof is conducted in three steps.

a) For any $x \in X$ there is an integer $n \in \mathbb{N}$ such that

$$x \in B_{\frac{n}{2}}(0) = nB_{\frac{1}{2}}(0).$$

Clearly, for any integer $n > 2\|x\|$, $x \in nB_{\frac{1}{2}}(0)$. It is therefore possible to cover X by such sets

$$X = \bigcup_{n \in \mathbb{N}} nB_{\frac{1}{2}}(0).$$

Since L maps X onto Y we can write

$$
\begin{aligned}
Y = L(X) &= L\left(\bigcup_{n\in\mathbb{N}} nB_{\frac{1}{2}}(0)\right) \\
&= \bigcup_{n\in\mathbb{N}} L\big(nB_{\frac{1}{2}}(0)\big).
\end{aligned}
$$

In fact, since the union above is equal to all of Y, we can take the closure of each set and obtain

$$
Y = \bigcup_{n\in\mathbb{N}} \overline{L\big(nB_{\frac{1}{2}}(0)\big)}. \tag{2.6}
$$

But Equation (2.6) expresses the complete space Y as the countable union of closed sets. By the Baire Category Theorem, Y cannot be the countable union of nowhere dense sets. Thus, there must be at least one integer $m \in \mathbb{N}$ such that

$$
\overline{L\big(mB_{\frac{1}{2}}(0)\big)}
$$

contains an open ball. But it is a simple matter to show that

$$
L\big(mB_{\frac{1}{2}}(0)\big) \equiv mL\big(B_{\frac{1}{2}}(0)\big).
$$

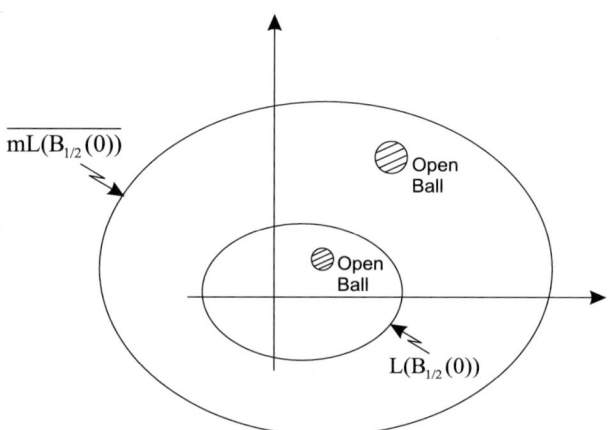

Figure 2.1: Schematic Diagram of Dilation of Open Ball

As depicted schematically in Figure 2.1 , if $\overline{L\big(mB_{\frac{1}{2}}(0)\big)}$ contains an open ball, so does $\overline{L\big(B_{\frac{1}{2}}(0)\big)}$.

b) Denote by $B_r(y_0)$ the open ball centered at y_0 with radius $r > 0$ that is contained in $\overline{L\big(B_{\frac{1}{2}}(0)\big)}$

$$
B_r(y_0) \subset \overline{L\big(B_{\frac{1}{2}}(0)\big)}. \tag{2.7}
$$

Of course, we can also write

$$B_r(0) \subset \overline{L\big(B_{\frac{1}{2}}(0)\big)} - y_0.$$

We want to show that

$$B_r(0) \subseteq \overline{L\big(B_{\frac{1}{2}}(0)\big)} - y_0 \subset \overline{L\big(B_1(0)\big)}. \tag{2.8}$$

Let ξ be some arbitrary element

$$\xi \in \overline{L\big(B_{\frac{1}{2}}(0)\big)} - y_0.$$

In this case we have

$$\xi + y_0 \in \overline{L\big(B_{\frac{1}{2}}(0)\big)}$$

and in view of equation

$$y_0 \in \overline{L\big(B_{\frac{1}{2}}(0)\big)}$$

we can find two sequences $\{\eta_k\}_{k\in\mathbb{N}}$ and $\{\gamma_k\}_{k\in\mathbb{N}}$, both contained in $L\big(B_{\frac{1}{2}}(0)\big)$, such that

$$\begin{aligned}
\eta_k &\rightarrow \xi + y_0 \\
\gamma_k &\rightarrow y_0
\end{aligned}$$

as $k \rightarrow \infty$. Since both sequences are contained in $L\big(B_{\frac{1}{2}}(0)\big)$, we know that there are pre-images $\{v_k\}_{k\in\mathbb{N}}$ and $\{w_k\}_{k\in\mathbb{N}}$ contained in $B_{\frac{1}{2}}(0)$ such that

$$\begin{aligned}
\eta_k &= Lv_k \\
\gamma_k &= Lw_k
\end{aligned}$$

for each $k \in \mathbb{N}$. But the difference $\{v_k - w_k\}_{k\in\mathbb{N}}$ is a sequence that is contained in $B_1(0)$

$$\|v_k - w_k\| \leq \|v_k\| + \|w_k\| < \frac{1}{2} + \frac{1}{2} = 1.$$

Moreover, the sequence $\{L(v_k - w_k)\}_{k\in\mathbb{N}}$ converges to ξ.

$$\begin{aligned}
\|\xi - L(v_k - w_k)\| &= \|\xi - Lv_k + Lw_k\| \\
&= \|\xi - \eta_k + \gamma_k\| \\
&\rightarrow \|\xi - (\xi + y_0) + y_0\| \\
&= 0
\end{aligned}$$

as $k \rightarrow \infty$. We have consequently shown that the inclusions shown in Equation (2.8) hold, since the point ξ was chosen arbitrarily in $\overline{L\big(B_{\frac{1}{2}}(0)\big)} - y_0$.

c) Now, we will use the inclusion

$$B_r(0) \subset \overline{L\big(B_1(0)\big)} \tag{2.9}$$

from Equation (2.8) to show that

$$B_{\frac{r}{2}}(0) \subset L\big(B_1(0)\big)$$

and the proof will be complete. First we observe from Equation (2.9) that

$$2^{-k}B_r(0) = B_{r2^{-k}}(0), \quad k \in \mathbb{N}$$

and

$$2^{-k}\overline{L\big(B_1(0)\big)} = \overline{L\big(B_{2^{-k}}(0)\big)}, \quad k \in \mathbb{N}.$$

That is, we can write

$$B_{r2^{-k}}(0) \subset \overline{L\big(B_{2^{-k}}(0)\big)} \tag{2.10}$$

for any integer $k \in \mathbb{N}$. Choose some arbitrary

$$z \in B_{\frac{r}{2}}(0).$$

But this means that the inclusion (2.10) holds for $k = 1$

$$z \in \overline{L\big(B_{2^{-1}}(0)\big)}.$$

There must be some element $z_1 = L\eta_1 \in L\big(B_{2^{-1}}(0)\big)$ such that

$$\|z - z_1\| = \|z - L\eta_1\| < \frac{r}{4} = r2^{-2} \tag{2.11}$$

where $\eta_1 \in B_{2^{-1}}(0)$. When we compare Equation (2.11) with the inclusion Equation (2.10), we see that Equation (2.10) holds for $k = 2$. That is

$$(z - L\eta_1) \in B_{r2^{-2}}(0) \subset \overline{L\big(B_{2^{-2}}(0)\big)}.$$

There must be a $z_2 = L\eta_2 \in L\big(B_{2^{-2}}(0)\big)$ such that

$$\|z - L\eta_1 - L\eta_2\| < \frac{r}{8} = r2^{-3}$$

where $\eta_2 \in B_{2^{-2}}(0)$. It is clear that we can proceed iteratively in this fashion. For each $k \in \mathbb{N}$,

$$\eta_k \in B_{2^{-k}}(0)$$

and the sequence $\{\sum_{k=1}^{n} \eta_k\}_{n \in \mathbb{N}}$ is a Cauchy sequence in a complete space. We must have

$$\lim_{n \to \infty} \sum_{k=1}^{n} \eta_k = \eta^* \in X.$$

Furthermore, $\eta^* \in B_1(0)$. This is true since

$$\left\| \sum_{k=1}^{\infty} \eta_k \right\| \leq \sum_{k=1}^{\infty} \|\eta_k\| < \sum_{k=1}^{\infty} \left(\frac{1}{2}\right)^k = 1.$$

Finally, we can also write

$$\left\| z - \sum_{k=1}^{n} L\eta_k \right\| < \frac{1}{2} r 2^{-n}.$$

We conclude that $L\eta^* = z$, and therefore

$$B_{\frac{r}{2}}(0) \subset L\big(B_1(0)\big). \qquad \qquad \square$$

Proof. Open Mapping Theorem

With the use of the last proposition, the proof of the Open Mapping Theorem now follows directly. Suppose that Q is an open set in X, and choose an arbitrary point $y = Lx \in LQ$. Since Q is open, there is an open ball $B_R(x)$ contained in Q where $R > 0$, centered at x. It follows immediately that

$$B_1(0) = \frac{1}{R}\big(B_R(x) - x\big) \subset \frac{1}{R}(Q - x).$$

By the previous proposition, the image of $B_1(0)$ contains an open ball θ about the origin $0 \in Y$.

We have

$$0 \in \theta \subseteq L\left(\frac{1}{R}(Q - x)\right)$$

$$0 \in \theta \subseteq \frac{1}{R}(LQ - Lx).$$

But this last inclusion implies that there is an open ball contained in LQ containing $y = Lx$. Since $y \in LQ$ was chosen arbitrarily, we conclude that LQ is open and the theorem is proved. $\qquad \qquad \square$

2.2   Dual Spaces

In the last few sections, we discussed how the set of all linear operators $L(X, Y)$ acting between any two normed vector spaces X and Y can be used to create the normed vector space of bounded linear operators , denoted by $\mathcal{L}(X, Y)$. Perhaps the most frequent use of the normed vector spaces $\mathcal{L}(X, Y)$ arises when Y is selected to be the real line $\mathbb{R}$, endowed with the usual topology. When we consider the space of all linear mappings from X into $\mathbb{R}$, we call this vector space the algebraic dual $X^\dagger$ of X. Each mapping $f \in X^\dagger$ is called a linear functional.

Definition 2.2.1. *A linear functional is a linear operator with domain in the vector space X and range in the scalar field of X. That is, if X is a real vector space, then*

$$f \colon \operatorname{dom}(f) \equiv X \to \mathbb{R} \text{ or } \mathbb{C}.$$

Definition 2.2.2. *A bounded linear functional f is a bounded linear operator with range in the scalar field of X, where $\operatorname{dom}(f) \equiv X$.*

Definition 2.2.3. *The algebraic dual space* $X^\dagger$ *of a vector space* X *is defined to be*

$$X^\dagger = L(X, \mathbb{R}).$$

When we further endow $L(X, \mathbb{R})$ with the operator topology defined in Definition 2.0.9, we obtain the topological dual space $X^* \triangleq \mathcal{L}(X, \mathbb{R})$ of X.

Definition 2.2.4. *Let* X *be a normed vector space. The set of all* bounded *linear functionals on* X *is a normed vector space with norm given by*

$$\|f\| = \sup_{x \neq 0} \frac{|f(x)|}{\|x\|} \quad \forall\, x \in X.$$

This space is called the topological dual space of X *and is denoted by* X^*.

We will only have occasion to use the topological dual in the remainder of this text; we consequently do not distinguish between the algebraic and topological dual space in our notation.

Each $f \in X^*$ is a bounded linear functional on X. We can write $f(x)$ to denote the action of f on a particular $x \in X$. Alternatively we use the notation

$$\langle f, x \rangle_{X^* \times X} = f(x)$$

to denote the action of f on x. The expression $\langle \cdot, \cdot \rangle_{X^*, X}$ is referred to as the duality pairing notation. This notation can be advantageous in applications with different dual spaces.

Important classes of linear operators can be defined in terms of the notion of duality. For example, the transpose or dual operators associated with a bounded linear operator will be used in numerous applications in this text.

Definition 2.2.5. *Let* U *and* V *be normed vector spaces and let* $A \in \mathcal{L}(U, V)$. *The transpose* $A' \in \mathcal{L}(V^*, U^*)$ *is the unique operator defined via the relationship*

$$\langle A'v', u \rangle_{U^* \times U} = \langle v', Au \rangle_{V^* \times V}$$

for all $v' \in V^*$ *and* $u \in U$.

The following theorem illustrates interesting and useful relationships between the nullspace and range of certain dual operators. These relationships are likewise used frequently in this book and make use of the annihilator of a given set.

Definition 2.2.6. *Let* X *be a normed vector space and* A *be a linear operator. Nullspace or kernel of operator* A *is a subspace* $\ker(A)$ *of* X *defined to be*

$$\ker(A) = \{x \in X : Ax = 0\}.$$

Definition 2.2.7. *Let* S *be a subset of a normed vector space* X. *The annihilator* $S^\perp$ *is defined to be the set of bounded linear functionals that evaluate to zero on every element of* S

$$S^\perp = \{A \in X^* : \langle A, x \rangle = 0 \quad \forall\, x \in S \subseteq X\}.$$

Now we can illustrate the usefulness of duality in characterizing how the range and nullspace of certain operators are related via the annihilator.

Theorem 2.2.1. *Let X and Y be normed vector spaces and suppose $A \in \mathcal{L}(X, Y)$. Then*

$$\big(\mathrm{range}(A)\big)^{\perp} = \ker(A').$$

Moreover, if the $\mathrm{range}(A')$ is closed, we have

$$\mathrm{range}(A') = \big(\ker(A)\big)^{\perp}.$$

Proof. Suppose $B \in \big(\mathrm{range}(A)\big)^{\perp}$. Then we have $\langle B, Ax \rangle_{Y^* \times Y} = 0$ for all $x \in X$. But this means $\langle A'B, x \rangle_{X^* \times X} = 0$ for all $x \in X$, which can be true only if $A'B \equiv 0$. Consequently $B \in \ker(A')$, and

$$\big(\mathrm{range}(A)\big)^{\perp} \subseteq \ker(A').$$

Suppose on the other hand, $z \in \ker(A')$. This means that $A'z = 0$. It follows that

$$\langle A'z, x \rangle_{X^* \times X} = \langle z, Ax \rangle_{Y^* \times Y} = 0 \qquad \text{for all } x \in X$$

which simply implies that $z \in \big(\mathrm{range}(A)\big)^{\perp}$. But this means that

$$\ker(A) \subseteq \big(\mathrm{range}(A)\big)^{\perp}.$$

Altogether we have

$$\big(\mathrm{range}(A)\big)^{\perp} = \ker(A'). \qquad \square$$

In the last theorem, we found that duality and the definition of the annihilator enabled a succinct relationship between the nullspace and range of certain dual operators. In the following theorem, we find that duality and the definition of the annihilator enables a characterization of the solution of an important minimum norm problem. This theorem will find use in various applications in this book.

Theorem 2.2.2. *Let U be a subspace of the real, normed vector space V. For some $v \in V$, define the distance*

$$d = \inf_{u \in U} \|u - v\|$$

from v to U. Then there is an element $A_0 \in U^{\perp} \subseteq V^*$ such that

$$d = \max_{\substack{\|A\| \leq 1 \\ A \in U^{\perp}}} \langle A, v \rangle_{V^* \times V} = \langle A_0, v \rangle_{V^* \times V}.$$

Proof. By the definition of the infimum, there is a sequence $\{u_k\}_{k \in \mathbb{N}} \subseteq U$ such that

$$\|u_k - v\| < d + \frac{1}{k}.$$

Choose any arbitrary $\tilde{A} \in U^\perp$ with $\|\tilde{A}\| = 1$. We can directly compute

$$\langle \tilde{A}, v \rangle = \langle \tilde{A}, v - u_k \rangle \leq \|\tilde{A}\| \, \|v - u_k\| < d + \frac{1}{k}.$$

Since k is arbitrarily large, we must have that

$$\langle \tilde{A}, v \rangle \leq d.$$

Consider the subspace $X \subseteq V$ that is defined as

$$X = \big\{ u + tv \,|\, u \in U, \, t \in \mathbb{R} \big\}$$

and define the functional $f : X \to \mathbb{R}$ such that

$$f(x) = td, \qquad x = u + tv, \ u \in U.$$

The functional f is linear on X. Suppose $x_1, \, x_2 \in X$ and

$$x_1 = u_1 + t_1 v, \ \ u_1 \in U$$
$$x_2 = u_2 + t_2 v, \ \ u_2 \in U$$

then

$$\begin{aligned} f(x_1 + x_2) &= (t_1 + t_2)d = f(x_1) + f(x_2) \\ f(\lambda x) &= (\lambda t)d = \lambda(td) = \lambda f(x). \end{aligned}$$

Let us show that the functional f has norm 1 over the subspace X. Let $x = tv + u$, where $u \in U$ and $t \in \mathbb{R}$. By definition, we have

$$\|f\| = \sup_{x \in X} \frac{|f(x)|}{\|x\|} = \sup_{u \in U} \frac{|td|}{\|tv + u\|} = \sup_{u \in U} \frac{|t|\,|d|}{|t| \, \left\|v + \frac{u}{t}\right\|} = \sup_{u \in U} \frac{d}{\left\|v + \frac{u}{t}\right\|}$$

$$= \frac{d}{\inf\limits_{u \in U} \left\|v + \frac{u}{t}\right\|} = \frac{d}{\inf\limits_{z \in U} \|v - z\|} \equiv 1.$$

By the Hahn-Banach theorem there is an extension of f to all of V having norm 1. Denote the extension by A_0. We have

$$\|A_0\| = 1$$
$$A_0\big|_X = f.$$

We claim that the extension $A_0 \in U^\perp$. Suppose $u \in U \subseteq X$ then it is easy to compute

$$\langle A_0, u \rangle = f(u) = 0 \cdot d = 0.$$

Moreover, $v \in X$, so that

$$\langle A_0, v \rangle = 1 \cdot d.$$

However, we have shown that $\langle \tilde{A}, v \rangle \leq d$ for any $\tilde{A} \in U^\perp$, and we have constructed $A_0 \in U^\perp$ such that $\langle A_0, v \rangle = d$. The theorem is proved. $\square$

There is simply an enormous literature that describes the structure of normed vector spaces, and their dual spaces. We cannot do justice to this theory in a short introduction. We would be remiss, however, if we did not address the role of dual spaces in the introduction of weak topologies on normed vector spaces. This topic plays an important role in numerous applications to mechanics.

2.3 The Weak Topology

Recall from Section 1.2 that it is possible, and often indispensable, that we are able to define different topologies on a given set X. Indeed, some of the most difficult questions in applied mathematics, control theory and mechanics arise in deciding which topology is appropriate for a given application. If a suitable topology can be found for a specific set X arising in an application, an enormous set of tools from functional analysis are then available for analyzing the intrinsic properties of that problem. Now, every normed vector space has an "intrinsic" topology derived from the norm. This topology is called the strong topology on X, and every open set in this topology is just the union of open balls $B_r(x)$ having the form

$$B_r(x) = \{y \in X : \|x - y\| < r\}.$$

If we unravel the notion of convergence of a net in this topology, we find that a sequence $\{x_n\}_{n \in \mathbb{N}}$ converges to $x \in X$ in the strong topology on X if for every $\epsilon > 0$, there is an $N(\epsilon) \in \mathbb{N}$ such that

$$k \geq N(\epsilon) \quad \Longrightarrow \quad x_k \in B_\epsilon(x).$$

Now, this can be a stringent requirement. If it is too difficult, or even impossible, to show that $x_k \to x$ strongly in X, it may be the case that we can define a topology with fewer neighborhoods, or open sets. Proving convergence can be considerably simpler in this coarser topology.

Simply stated, one of the most frequently employed, alternative topologies that we can consider is the *weak topology from* X^*. Recall from the definition of weak topology that any collection of continuous functions from one topological space (X, τ_X) to another (Y, τ_Y) can be used to define a new topology on X. By convention, if we choose the space (Y, τ_Y) to be $(\mathbb{R}, \text{Euclidean})$ and consider all of the continuous linear functionals from (X, τ_X) into $(\mathbb{R}, \text{Euclidean})$ the resulting weak topology is called "the" weak topology on X. It is a simple exercise to show that convergence in the weak topology is characterized by the following theorem:

Theorem 2.3.1. *Let X be a normed vector space. A sequence $\{x_k\}_{k \in \mathbb{N}}$ in X is convergent in the weak topology on X if*

$$f(x_k) \to f(x) \ \ in \ \mathbb{R} \ \forall f \in X^*.$$

As noted in the construction of general weak topologies in Section 1.2, the weak topology on X is generated by the set of finite intersections of inverse images of open sets in $\mathbb{R}$ under continuous functionals in X^*. A local base at $x_0 \in X$ for the weak topology is then given by

$$B_r(x_0; f_1, \ldots, f_n) \triangleq \bigcap_{f \in \mathcal{F}_n} \left\{ x \in X : |\langle f, x - x_0 \rangle| < r \right\}$$

where $\mathcal{F}_n = \{f_1, \ldots, f_n\}$ is any collection of n functionals contained in X^* and $r > 0$.

2.4   The Weak* Topology

So we have seen that given a normed vector space X, there is a useful, direct construction of a topology on X that is weaker than the norm topology. We take the set of all bounded, linear functionals, endow this set with the "usual" algebraic structure, further endow it with the operator norm, and we get the topological dual X^*. "The weak topology" on X is the weak topology in the sense of Definition 1.2.16. But we do not need to stop at this point. We can carry out the same construction starting with X^*. This process gives rise to the bidual, or second dual space, X^{**}.

Definition 2.4.1. *Let X be a normed vector space. The second dual space, or bidual, of the space X is X^{**}.*

Now at first this might seem like a sterile mathematical exercise. If we are truly interested in X, as being descriptive of a *physical system*, it is relatively clear that the construction of X^* is interesting in that it describes alternate characterizations of convergence of states in our *physical system*. What is the use in constructing X^{**} that is, strictly speaking, a way of defining a topology on a space that is "one space away" from our physical space X?

The answer is that there is an interesting and useful relationship between X and X^{**}: we can view each element of $x \in X$ "as an element" of X^{**}. Now to see how we can make this identification, pick an arbitrary pair of elements $x, y \in X$. By the definition of X^*, if $f \in X^*$, we know that

$$\begin{aligned} f(x + y) &= f(x) + f(y) \\ f(\lambda x) &= \lambda f(x). \end{aligned}$$

We can define two mappings g and h that act on X^*

$$\begin{aligned} g(f) &= f(x) \\ h(f) &= f(y). \end{aligned}$$

Clearly, g and h are determined by the fixed elements x and y from X. If we define addition of g and h to be

$$(g + h)(f) = g(f) + h(f) \quad \forall f \in X^*$$

then a moment's reflection shows that the set of all functions defined via this construction on X^* is a vector space. That is,

$$S = \{g \in \mathcal{L}(X^*, \mathbb{R}) \mid \exists x \in X : g(f) = f(x) \; \forall f \in X^*\}$$

is a vector space, and each of its elements is uniquely determined by an element of X. In fact, much more is true. We can show that the above construction defines a mapping $\mathcal{I} : X \to S \subseteq X^{**}$. This mapping is called the canonical injection of X into X^{**}.

Theorem 2.4.1. *Let X be a normed vector space. There exists a canonical injection $\mathcal{I}$ of X into its bidual X^{**}*

$$\mathcal{I} : X \to X^{**}$$

such that

$$\mathcal{I} : X \to \text{range}(\mathcal{I}) \subseteq X^{**}$$

is a linear isometry.

Thus, Theorem 2.4.1 has shown that a normed vector space can be "identified" with a subset of X^{**}. In the case when the range of the canonical injection $\mathcal{I}$ is all of X^{**}, that is $\mathcal{I}$ is onto, X is said to be a reflexive normed vector space.

Definition 2.4.2. *A normed vector space X is reflexive if the canonical injection $\mathcal{I} : X \to X^{**}$ is onto. That is,*

$$\text{range}(\mathcal{I}) \equiv X^{**}.$$

Having defined the canonical injection

$$\mathcal{I} : X \mapsto X^{**}$$

it is possible to define yet another weak topology in the sense introduced in Definitions 1.2.17 and 1.2.18 for general topological spaces. The collection of functions

$$\text{range}\,(\mathcal{I}) \subseteq X^{**}$$

defines a collection of bounded linear functionals in X^{**}. We can use this collection of functions to define a weak topology on X^*. This topology is denoted the weak* topology on X^*

$$(X^*, \text{weak}^*) = (X^*, \text{weak}(\text{range}\,\mathcal{I})).$$

As in Theorem 2.3.1, there is direct characterization of convergent sequences in X^* in the $(X^*, weak^*)$ topology.

Theorem 2.4.2. *Let X be a normed vector space. A sequence of functionals $\{f_n\}_{n \in \mathbb{N}} \subseteq X^*$ converges to $f \in X^*$ in the weak* topology if*

$$f_n(x) \to f(x) \qquad \forall\, x \in X.$$

Proof. The proof is based on the general definition of weak* convergence in a topological space. From Theorem 1.2.4, we know that

$$f_n \to f$$

in (X^*, weak^*) if and only if

$$g(f_n) \to g(f)$$

for all $g \in \text{range}(\mathcal{I}) \subseteq X^{**}$. If we unravel the definition of the canonical embedding $\mathcal{I}$, we obtain the conclusion of the theorem. □

Alternatively, a local base at $\eta \in X^*$ for the weak* topology is given by the open sets

$$B_r(\eta; x_1, \ldots, x_n) = \bigcap_{x \in X_n} \left\{ f \in X^* : |\langle f - \eta, x \rangle| < r \right\}$$

where $X_n = \{x_1, \ldots, x_n\}$ is any collection of n elements in X and $r > 0$.

The usefulness of the weak* topology in applications is immediately apparent. One of the most pragmatic properties of the weak* topology is that it contains a large collection of compact sets. As the following theorem shows, the closed unit ball of the dual space of a Banach space is weak* compact. While the proof of the following theorem is beyond the scope of this text (see for example [3]), it is used frequently in computational methods and mechanics.

Theorem 2.4.3 (Alaoglu's Theorem). *The closed unit ball of the dual space X^* of a normed space X is compact in the weak* topology.*

Theorem 2.4.3 is of a very general nature. Various approaches exist for its proof, as typified in [20, p. 229] or [30, p. 174]. More specialized results, that assert the relative compactness of sequences under certain separability or reflexivity assumptions may find more direct application for engineers and scientists.

Theorem 2.4.4. *Suppose X is a separable normed linear space. Every bounded sequence in X^* contains a subsequence that is weak* convergent.*

Proof. Suppose that $\{f_k\}_{k \in \mathbb{N}}$ is a bounded sequence in X^* and that $\{x_k\}_{k \in \mathbb{N}}$ is a dense collection of vectors in X. We construct a convergent subsequence $\{f_{k,k}\}_{k \in \mathbb{N}} \subseteq \{f_k\}_{k \in \mathbb{N}}$ via a diagonalization argument. First consider the sequence of real numbers

$$|\langle f_k, x_1 \rangle| \leq \|f_k\| \, \|x_1\| < C_1$$

that is bounded in $\mathbb{R}$. We can extract a subsequence

$$\{f_{k,1}\}_{k \in \mathbb{N}} \subseteq \{f_k\}_{k \in \mathbb{N}}$$

such that the sequence of real numbers

$$\left\{ \langle f_{k,1}, x_1 \rangle \right\}_{k \in \mathbb{N}}$$

converges in $\mathbb{R}$. Subsequently, we note that the sequence of real numbers

$$\left|\langle f_{k,1}, x_2 \rangle\right| \leq \|f_{k,1}\| \, \|x_2\| < C_2$$

is bounded. We can extract a further subsequence

$$\{f_{k,2}\}_{k \in \mathbb{N}} \subseteq \{f_{k,1}\}_{k \in \mathbb{N}} \subseteq \{f_k\}_{k \in \mathbb{N}}$$

such that the sequence of real numbers

$$\left\{\langle f_{k,2}, x_1 \rangle\right\}_{k \in \mathbb{N}}$$
$$\left\{\langle f_{k,2}, x_2 \rangle\right\}_{k \in \mathbb{N}}$$

converge to real numbers. Proceeding in this fashion, we obtain a diagonal sequence

$$\{f_{k,k}\}_{k \in \mathbb{N}} \subseteq \{f_k\}_{k \in \mathbb{N}}$$

such that the sequence of real numbers

$$\left\{\langle f_{k,k}, x_j \rangle\right\}_{k \in \mathbb{N}}$$

converges for any fixed $j \in \mathbb{N}$. Fix $x \in X$. By the triangle inequality we have

$$\left|\langle f_{j,j}, x \rangle - \langle f_{k,k}, x \rangle\right| \leq \left|\langle f_{j,j}, x \rangle - \langle f_{j,j}, x_m \rangle\right| + \left|\langle f_{j,j}, x_m \rangle - \langle f_{k,k}, x_m \rangle\right|$$
$$+ \left|\langle f_{k,k}, x_m \rangle - \langle f_{k,k}, x \rangle\right|$$
$$\leq 2C\|x - x_m\| + \left|\langle f_{j,j} - f_{k,k}, x_m \rangle\right|.$$

Fix $\epsilon > 0$. Since $\{x_m\}_{m \in \mathbb{N}}$ is dense in X, we can choose m large enough so that

$$\|x - x_m\| < \frac{1}{3C}\epsilon.$$

Furthermore, for m so selected, there is an N such that $j, k > N$ implies

$$\left|\langle f_{j,j} - f_{k,k}, x_m \rangle\right| < \frac{1}{3}\epsilon.$$

But this means that $j, k > N$ implies

$$\left|\langle f_{j,j}, x \rangle - \langle f_{k,k}, x \rangle\right| < \epsilon$$

and the sequence

$$\left\{\langle f_{j,j}, x \rangle\right\}_{j \in \mathbb{N}}$$

is consequently a Cauchy sequence of real numbers, and therefore converges to a real number we denote by $\langle f, x \rangle$. The assignment so defined is linear and bounded, hence $f \in X^*$. $\qquad\square$

This theorem can, in fact, lead to another characterization of compact sequences, in this case for bounded sequences in reflexive normed spaces.

Theorem 2.4.5. *Let X be a normed vector space. X is reflexive if and only if every bounded sequence in X contains a weakly convergent subsequence.*

Proof. For this text, we will only use the implication that if a normed space is reflexive then a bounded sequence in this space contains a convergent subsequence. The reverse implication is proven in [20, p. 119]. Suppose $\{x_k\}_{k\in\mathbb{N}}$ is a bounded sequence in the reflexive normed vector space X. Then the subspace

$$V \triangleq \overline{\operatorname{span}\{x_k\}_{k\in\mathbb{N}}} \subseteq X$$

is separable and reflexive. V is reflexive since any closed subspace of a reflexive Banach space is reflexive. Moreover, the construction of the canonical isometry of V onto V^{**} guarantees that V^{**} is separable. If Y^* is separable, then Y is also separable for any normed vector space Y and its dual Y^*. Hence, V^* is separable. We know that $\{\mathcal{I}(x_k)\}_{k\in\mathbb{N}} \subseteq V^{**}$ is bounded since $\mathcal{I}$ is an isometric mapping of V onto V^{**}. By Theorem 2.4.4, there is a subsequence $\{\mathcal{I}(x_{k_j})\}_{k_j\in\mathbb{N}}$ that is weak* convergent to some $\mathcal{I}(x) \in V^{**}$. Since $\mathcal{I}$ is a linear isometry that is onto V^{**}, the sequence $\{x_{k_j}\}_{k_j\in\mathbb{N}}$ is weakly convergent to x in V. In general, we have the inclusions

$$V \subseteq X \text{ and } X^* \subseteq V^*.$$

We conclude that the sequence $\{x_{k_j}\}_{k_j\in\mathbb{N}}$ is in fact convergent to x in the weak topology of X. $\square$

We close this section by considering a very specific interpretation of the weak* topology restricted to the closed unit ball of the dual space X^* of a separable normed vector space. In fact, we can show that the relative weak* topology that the closed unit ball of dual space inherits as a subset of X^* is metrizable. This metric in induced by a norm called the weak norm on X^*.

Theorem 2.4.1. *Let X be a separable normed vector space and let S_{X^*} be the closed unit ball of the dual space X^**

$$S_{X^*} = \{f \in X^* : \|f\|_{X^*} \leq 1\}.$$

(1) *If $\{x_k\}_{k\in\mathbb{N}}$ is dense in X, the expression*

$$\|f\|_w \triangleq \sum_{k=1}^{\infty} 2^{-k} \frac{|f(x_k)|}{(1 + \|x_k\|)}$$

defines a norm on X^*. This norm is referred to as the weak norm on X^*.

(2) *The relativization of the weak norm topology on X^* induced by $\|\cdot\|_w$ and the relativization of the weak* topology on X^* coincide on S_{X^*}.*

Proof (1). By inspection, we see that

$$\|\alpha f\|_w = \sum_{k=1}^{\infty} 2^{-k} \frac{|\alpha f(x_k)|}{(1 + \|x_k\|)}$$

$$= |\alpha| \sum_{k=1}^{\infty} 2^{-k} \frac{|f(x_k)|}{(1 + \|x_k\|)} = |\alpha| \, \|f\|_w.$$

Likewise, the function $\| \cdot \|_w$ satisfies a triangle inequality

$$\|f + g\|_w = \sum_{k=1}^{\infty} 2^{-k} \frac{|f(x_k) + g(x_k)|}{(1 + \|x_k\|)}$$

$$\leq \sum_{k=1}^{\infty} 2^{-k} \frac{\left\{ |f(x_k)| + |g(x_k)| \right\}}{(1 + \|x_k\|)}$$

$$= \|f\|_w + \|g\|_w.$$

We would like to show that

$$\|f\|_w = 0 \quad \Longrightarrow \quad f - 0 \quad \text{in} \quad X^*.$$

From the definition of the weak norm, we see that

$$\|f\|_w = \sum_{k=1}^{\infty} 2^{-k} \frac{|f(x_k)|}{1 + \|x_k\|} = 0 \quad \Longrightarrow \quad |f(x_k)| = 0 \quad \forall k \in \mathbb{N}.$$

Let $x \in X$ be arbitrary. For any such x there is a subsequence $\{x_{k_j}\}_{k_j \in \mathbb{N}} \subseteq \{x_k\}_{k \in \mathbb{N}}$ such that

$$x_{k_j} \to x.$$

By the continuity of f, we must have

$$f(x_{k_j}) \to f(x).$$

Since each $f(x_{k_j}) = 0$, we have

$$f(x) = 0 \quad \forall x \in X.$$

That is,

$$\|f\|_w = 0 \quad \Longrightarrow \quad f = 0 \quad \text{in} \quad X^*.$$

Hence $\| \cdot \|_w$ is a norm on X^*. $\qquad \square$

Proof (2). Recall that in this theorem we consider two topologies on X^*. Denote by $(X^*, \tau_{\|\cdot\|_w})$ the topology on X^* induced by the weak norm $\| \cdot \|_w$, and by (X^*, τ_{w^*})

the weak* topology on X^*. A base for $(X^*, \tau_{\|\cdot\|_w})$ is given by the collection of $\tau_{\|\cdot\|_w}$-open sets

$$B_r(f) = \{g \in X^* : \|g - f\|_w < r\} \qquad f \in X^*, \ r > 0$$

indexed by the center f of the ball and radius r. A base for the space (X^*, τ_{w^*}) is the collection of τ_{w^*}-open sets

$$B_r(f; \xi_1, \xi_2, \ldots, \xi_n) = \left\{g \in X^* : \max_{i=1,\ldots,n} |g(\xi_i) - f(\xi_i)| < r\right\}$$

indexed by the center f, the finite integer n, points $\xi_1, \ldots, \xi_n \in X$, and radius r. Now, suppose that θ_{w^*} is an open set in the relativization of the weak* topology (X^*, τ_{w^*}) to S_{X^*}. What does this mean? Pick any point $f \in \theta_{w^*}$. By the definition of the base for the weak* topology, there is a finite number n, a set of points $\xi_1, \ldots, \xi_n$, and a real number $r > 0$ such that

$$S_{X^*} \cap B_r(f; \xi_1, \xi_2, \ldots, \xi_n) \subseteq \theta_{w^*}. \tag{2.12}$$

That is, we can find one of the sets in the base for (X^*, τ_{w^*}) relativized to S_{X^*} that contains f and is "small enough" to be contained in θ_{w^*}. We first note that the sequence $\{x_k\}_{k \in \mathbb{N}}$ is dense in X by definition. Thus, it is always possible to find candidates

$$\{x_{k_1}, x_{k_2}, \ldots, x_{k_n}\} \subseteq \{x_k\}_{k \in \mathbb{N}} \tag{2.13}$$

that approximate the fixed points $\{\xi_1, \xi_2, \ldots, \xi_n\}$ in Equation (2.12) to any desired degree of accuracy. For example, we can choose the candidates in Equation (2.13) such that

$$\|x_{k_j} - \xi_j\| \le \epsilon r \qquad j = 1, \ldots, n \tag{2.14}$$

where $\epsilon < \frac{1}{3}$. The choice of approximation in equation (2.14) determines a set of integers $\{k_1, k_2, \ldots, k_n\}$ that picks out certain terms in the sequence $\{x_k\}_{k \in \mathbb{N}}$. Define the constant

$$C \overset{\triangle}{=} \inf_{j=1,\ldots,n} \left\{ \frac{2^{-k_j}}{1 + \|x_{k_j}\|} \right\}.$$

Now, suppose that

$$g \in S_{X^*} \cap B_{\tilde{r}}(f)$$

where

$$\tilde{r} \overset{\triangle}{=} \epsilon C \cdot r.$$

In this case, we can write

$$\|f - g\|_w \le \epsilon C \cdot r. \tag{2.15}$$

But the left-hand side of inequality (2.15) can be bounded below as follows

$$
\begin{aligned}
\|f - g\|_w &= \lim_{n \to \infty} \sum_{j=1}^{n} 2^{-j} \frac{|(f-g)x_j|}{1 + \|x_j\|} \\
&\geq \sum_{j=1}^{n} 2^{-k_j} \frac{|(f-g)x_{k_j}|}{1 + \|x_{k_j}\|} \\
&\geq \inf_{j=1,\ldots,n} \left\{ \frac{2^{-k_j}}{1 + \|x_{k_j}\|} \right\} \sum_{j=1}^{n} |(f-g)x_{k_j}| \\
&= C \sum_{j=1}^{n} |(f-g)x_{k_j}|. \tag{2.16}
\end{aligned}
$$

Combining inequalities (2.15) and (2.16), we see that

$$
|(f-g)x_{k_j}| \leq \epsilon r.
$$

Finally, we can write

$$
\begin{aligned}
|f(\xi_j) - g(\xi_j)| &= |(f-g)(\xi_j)| \\
&\leq |(f-g)(x_{k_j}) - (f-g)(x_{k_j} - \xi_j)| \\
&\leq \epsilon r + (\|f\|_{X^*} + \|g\|_{X^*})\|x_{k_j} - \xi_j\| \\
&\leq \epsilon r + 2 \cdot \epsilon r = 3\epsilon r < r. \tag{2.17}
\end{aligned}
$$

In this last inequality we have used the fact that

$$
f,\, g \in S_{X^*}.
$$

But Equation (2.17) implies that

$$
\left(S_{X^*} \cap B_{\bar{r}}(f) \right) \subseteq \left(S_{X^*} \cap B_r(f; \xi_1, \ldots, \xi_n) \right) \subseteq \theta_{w^*}.
$$

This process can be repeated for any choice of $f \in \theta_{w^*}$. Hence, θ_{w^*} is the union of sets extracted from a basis for the relativization of the weak norm topology to S_{X^*}. We can conclude that θ_{w^*} is an open set in the relativized weak norm topology.

Now suppose that θ_w is an open set in the relativization of the weak norm topology (X^*, τ_w) to S_{X^*}. Choose any $f \in \theta_w$. By the definition of a basis for this topology, there is an $r > 0$ such that

$$
S_{X^*} \cap B_r(f) \subseteq \theta_w. \tag{2.18}
$$

Recall that the inclusion (2.18) implies that for any $g \in B_r(f)$, we have

$$
\lim_{n \to \infty} \sum_{j=1}^{n} 2^{-j} \frac{|(f-g)x_j|}{1 + \|x_j\|} \leq r.
$$

Of course, the summation above, being convergent, implies that the tail of the series converges to zero. There is an index $K \in \mathbb{N}$, consequently, such that

$$\lim_{n \to \infty} \sum_{j=K+1}^{n} 2^{-j} \frac{|(f-g)x_j|}{1+\|x_j\|} \leq \epsilon r.$$

Define the constant

$$C \triangleq \sup_{j=1,\,\ldots,\,K} \frac{2^{-j}}{1+\|x_j\|}.$$

Now, suppose that $g \in S_{X^*}$ and

$$|f(x_j) - g(x_j)| \leq \epsilon \frac{r}{C \cdot K}, \qquad j = 1, \ldots, K \tag{2.19}$$

under these conditions, we can bound $|f - g|_w$ via a direct computation.

$$
\begin{aligned}
|f-g|_w &= \lim_{n \to \infty} \sum_{k=1}^{n} 2^{-k} \frac{|(f-g)x_k|}{1+\|x_k\|} \\
&\leq \underbrace{\sum_{k=1}^{K} 2^{-k} \frac{|(f-g)x_k|}{1+\|x_k\|}}_{\leq C \sum_{k=1}^{K} |(f-g)x_k|} + \underbrace{\lim_{n \to \infty} \sum_{k=K+1}^{n} 2^{-k} \frac{|(f-g)x_k|}{1+\|x_k\|}}_{\leq \epsilon r}.
\end{aligned}
$$

But from inequality (2.19), we conclude that

$$|f-g|_w \leq C \sum_{k=1}^{K} \epsilon \frac{r}{C \cdot K} + \epsilon r = 2\epsilon r < r.$$

Thus, we have shown that

$$S_{X^*} \cap B_r(f; x_1, \ldots, x_K) \subseteq S_{X^*} \cap B_r(f) \subseteq \theta_w.$$

This argument can be repeated for every $f \in \theta_w$. Consequently, θ_w is the union of basis sets for the relativization of the weak* topology to S_{X^*}. The set θ_w is thereby open in this topology. $\qquad \square$

2.5   Signed Measures and Topology

In the previous chapter, we discussed the rudiments of measure theory, and its role in the development of integration theory. In several of the examples that follow in this book, it is important that some classes of measures can be interpreted as a subspace of the dual space of a Banach space. We have defined a measure as an extended, real-valued set function

$$\mu : A \longrightarrow [0, +\infty] \subset \overline{\mathbb{R}}, \qquad A \in \mathcal{B}$$

where $\mathcal{B}$ is a σ-algebra of subsets of the underlying set X. By definition, a measure is countably additive on the σ-algebra $\mathcal{B}$ of X, and the measure of the empty set is always zero. The pair consisting of the set X and a σ-algebra of subsets of X is called a measurable space. Perhaps the most direct method for introducing a topology on measures is to define the set of *signed measures* associated with a given measurable space $(X, \mathcal{B})$.

Definition 2.5.1. *A signed measure μ defined on the measurable space $(X, \mathcal{B})$ is an extended real-valued set function such that*

  (i) *The set function μ assumes at most one of the values $+\infty, -\infty$,*
 (ii) *the measure of the empty set is zero, $\mu(\emptyset) = 0$, and*
(iii) *the set function μ is countably additive for any sequence of pairwise disjoint sets extracted from $\mathcal{B}$.*

Some care must be taken in the interpretation of property (iii) above. If $\mu\left(\bigcup_i A_i\right)$ is finite, then

$$\mu\left(\bigcup_i A_i\right) = \sum_i \mu(A_i) \tag{2.20}$$

implies that the series in Equation (2.20) converges absolutely.

It is essential in many applications that every signed measure can be decomposed in a "canonical fashion" into measures. While the axiomatic development of the "canonical decomposition" of signed measures is somewhat lengthy, the idea is straightforward. The reader is referred to [6, 9, 27] for the details; we simply outline the procedure here for completeness. Roughly speaking, we restrict the signed measure to a particular collection of subsets of $\mathcal{B}$ having a positive signed measure, and subsequently define a new measure. Note that we cannot choose just *any* collection of sets having positive signed measure in this construction, as some sets might contain subsets that have negative signed measure. Likewise, we restrict the signed measure to a carefully selected collection of subsets in $\mathcal{B}$ that have negative signed measure, take the negative of this set function, and define another measure. We consequently achieve a decomposition of a measure that is analogous to the decomposition of a real-valued function into its positive and negative components. To precisely describe which subsets in $\mathcal{B}$ are used to define the restricted measures, we must introduce the notion of positive and negative sets for a signed measure. If μ is a signed measure on the measurable space $(X, \mathcal{B})$, a subset $A \in \mathcal{B}$ is positive for μ if $\mu(B) \geq 0$ for every measurable set $B \subseteq A$. We define a negative set for μ similarly. It is a remarkable fact that for any signed measure μ, there are disjoint subsets $A, B \in \mathcal{B}$ such that

$$X = A \cup B$$

where A is positive for μ and B is negative for μ. This representation of the set X is known to be *Hahn decomposition*, which, in general, is not unique. Associated to such a distinguished decomposition of X, we can construct our pair of measures μ^+ and μ^-, denoted the positive and negative measures corresponding to μ

$$
\begin{aligned}
\mu^+(C) &= \mu(A \cap C) \quad \forall C \in \mathcal{B} \\
\mu^-(C) &= -\mu(B \cap C) \quad \forall C \in \mathcal{B}.
\end{aligned}
$$

Since A and B are disjoint, we have in particular that

$$
\mu^+(B) = 0
$$
$$
\mu^-(A) = 0.
$$

A pair of measures μ_1 and μ_2 are said to be mutually singular if there are disjoint measurable sets such that $X = A \cup B$ and $\mu_1(B) = \mu_2(A) = 0$. We have the following proposition guaranteeing the uniqueness of this decomposition:

Proposition 2.5.1. *Let μ be a signed measure on the measurable space $(X, \mathcal{B})$. Then there are two mutually singular measures μ^+ and μ^- on $(X, \mathcal{B})$ such that*

$$
\mu = \mu^+ - \mu^-.
$$

Moreover, there is only one such pair of measures. The representation of μ by μ^+ and μ^- is called Jordan decomposition.

The total variation of a signed measure μ is subsequently defined as

$$
|\mu|(C) = \mu^+(C) + \mu^-(C).
$$

We now have arrived at our goal; we can define the precise sense in which measures can be endowed with a topology.

Theorem 2.5.1. *Let X be a compact metric space and denote by $C(X)$ the Banach space of all real-valued, continuous functions on X endowed with the usual sup-norm*

$$
\|f\|_{C(X)} = \sup_{x \in X} |f(x)|.
$$

The set of all finite, signed Borel measures on X is isometrically isomorphic to the dual space of $C(X)$. The action of each μ on $f \in C(X)$ is given by the integration formula

$$
\mu(f) = \int_X f \, d\mu \triangleq F(f)
$$

where $F \in C(X)^*$. The mapping $\mu \mapsto F$ defines the isometric isomorphism where

$$
|\mu|(X) \equiv \|F\|_{C(X)^*}.
$$

The structure of these spaces is depicted graphically in Figure 2.2.

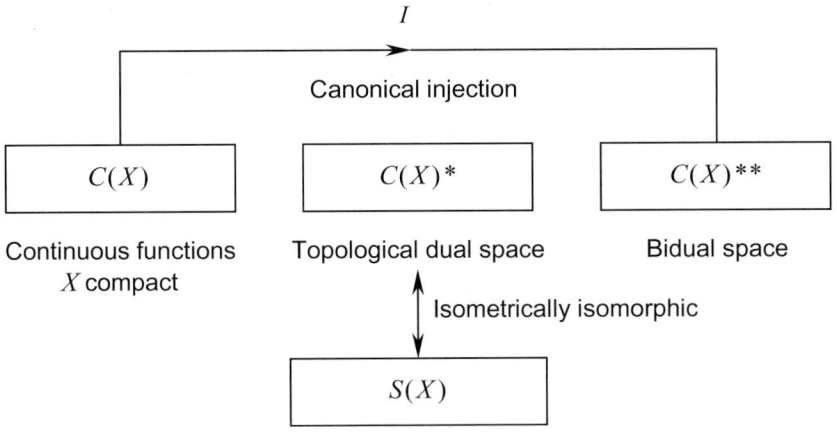

Figure 2.2: $C(X)$, its Dual and Bidual

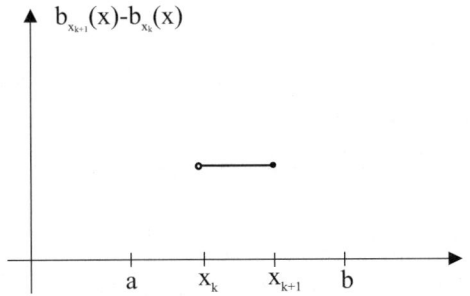

Figure 2.3: Difference adjacent basis functions

2.6 Riesz's Representation Theorem

A fundamental issue in Operator Theory is the representation of bounded linear functionals on different normed vector spaces. Theorem 2.5.1 describes one such representation. This theorem is sometimes known as Riesz's Representation Theorem for continuous functions.

2.6.1 Space of Lebesgue Measurable Functions

The proof of Theorem 2.5.1 on a general compact metric space can be quite lengthy. Detailed proofs are given in [3, 28]. However, it is possible to obtain a constructive proof of Riesz's Representation Theorem in the special case that the compact set $X = [a, b] \subseteq \mathbb{R}$. In this case we can see that the dual of $C[a, b]$ can be identified in terms of certain functions of bounded variation.

Theorem 2.6.1 (Riesz's Representation Theorem). *Let f be a bounded linear functional on $C[a, b]$. Then there is a function g of bounded variation such that for all $h \in C[a, b]$*

$$f(h) = \int_a^b h(x) dg(x) \tag{2.21}$$

and

$$\|f\|_{C^*[a,b]} = V(g). \tag{2.22}$$

Conversely, every function g of bounded variation on $[a, b]$ defines a bounded linear functional on $C[a, b]$ via (2.21) that satisfies (2.22).

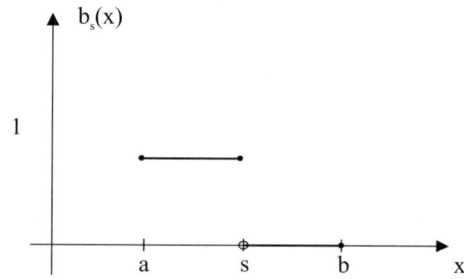

Figure 2.4: Basis function for Theorem 2.6.1

Proof. Suppose f is a bounded linear functional on $C[a, b]$. The space $C[a, b]$ is a subspace of the space $B[a, b]$ of all bounded functions defined on $[a, b]$. By the Hahn-Banach theorem, there is an extension F of f to all of $B[a, b]$ such that

$$\|F\| = \|f\|_{C^*[a,b]}.$$

We define the step functions $b_s \in B[a, b]$ for each $s \in [a, b]$ according to

$$b_s(x) = \begin{cases} 1, & a \le x \le s, \\ 0, & s < x \le b. \end{cases}$$

This function is depicted schematically in Figure 2.4. We claim that the function g appearing in the representation formula in (2.21) is simply

$$g(s) = F(b_s).$$

This expression makes sense since $b_s \in B[a, b]$ and F is a bounded linear functional on $B[a, b]$. We need to show that g is of bounded variation, the integral representation in (2.21) holds, and the norm of the bounded linear functional f

is equal to the variation of g. Let us show that g has bounded variation. Let $a = x_0 < x_1 < \cdots < x_n = b$ be a partition of $[a, b]$. We can compute

$$\sum_{k=0}^{n-1} \left| g(x_{k+1}) - g(x_k) \right| = \sum_{k=0}^{n-1} \left| F(b_{x_{k+1}}) - F(b_{x_k}) \right|$$

$$= \sum_{k=0}^{n-1} \lambda_k \big(F(b_{x_{k+1}}) - F(b_{x_k}) \big)$$

where

$$\lambda_k = \text{sign}\big(F(b_{x_{k+1}}) - F(b_{x_k}) \big).$$

By the linearity and boundedness of F on $B[a, b]$, we have

$$\sum_{k=0}^{n-1} \left| g(x_{k+1}) - g(x_k) \right| = F\left(\sum_{k=0}^{n-1} \lambda_k (b_{x_{k+1}} - b_{x_k}) \right)$$

$$\leq \|F\| \left\| \sum_{k=0}^{n-1} \lambda_k (b_{x_{k+1}} - b_{x_k}) \right\|.$$

But since

$$b_{x_{k+1}}(x) - b_{x_k}(x) = \begin{cases} 1, & x_k < x \leq x_{k+1} \\ 0, & \text{otherwise} \end{cases}$$

and $|\lambda_k| = 1$, we have

$$\left\| \sum_{k=0}^{n-1} \lambda_k (b_{x_{k+1}} - b_{x_k}) \right\| = \sup_{x \in [a,b]} \left| \lambda_k \big(b_{x_{k+1}}(x) - b_{x_k}(x) \big) \right| = 1.$$

We can conclude that

$$\sum_{k=0}^{n-1} \left| g(x_{k+1}) - g(x_k) \right| \leq \|F\| = \|f\|. \tag{2.23}$$

Thus $V(g) \leq \|f\|$. Next, we show that the integral representation in (2.21) holds. Let $h \in C[a, b]$. Define

$$r_n(x) \triangleq \sum_{k=0}^{n-1} h(x_k) \big(b_{x_{k+1}}(x) - b_{x_k}(x) \big).$$

We have

$$\|h - r_n\|_{B[a,b]} = \sup_{x \in [a,b]} \left| h(x) - r_n(x) \right|$$

$$= \max_{k=0, \ldots, n-1} \; \max_{x_k \leq x \leq x_{k+1}} \left| h(x) - h(x_k) \right|.$$

By the uniform continuity of $h(x)$, which is a continuous function on a compact set, we have

$$r_n \to h \quad \text{in} \quad B[a,b]$$

as $n \to \infty$. But F is a continuous functional on $B[a,b]$ so that

$$F(r_n) \to F(h)$$

as $n \to \infty$. Since $h \in C[a,b]$

$$F(h) = f(h).$$

Moreover

$$F(r_n) = F\left(\sum_{k=0}^{n-1} h(x_k)(b_{x_{k+1}} - b_{x_k}) \right)$$

$$= \sum_{k=0}^{n-1} h(x_k)\big(g(x_{k+1}) - g(x_k)\big)$$

$$\to \int_a^b h(x)dg(x).$$

Thus, we have

$$f(h) = \int_a^b h(x)dg(x).$$

Finally we can calculate

$$|f(h)| = \left| \int_a^b h(x)dg(x) \right| \leq \|h\|V(g).$$

So that

$$\frac{|f(h)|}{\|h\|} \leq V(g).$$

But this implies that

$$\|f\| \leq V(g).$$

Combining this result with the conclusions from (2.23), we have

$$\|f\| = V(g). \qquad \qquad \square$$

2.6.2 Hilbert Spaces

The introduction of the notion of "orthogonality" in inner product spaces adds considerably more structure than normed vector spaces. One practical implication is that the class of bounded linear functionals on Hilbert spaces have an explicit representation. Each bounded linear functional defined on a Hilbert space can be represented in terms of an inner product with a distinguished, or representative, element of H. This celebrated result is known as Riesz's Representation Theorem for bounded linear functionals on Hilbert Spaces.

Theorem 2.6.2 (Riesz's Representation Theorem). *Let H be a Hilbert space, and let $f \in H^*$, that is, f is a bounded linear functional on H. Then, the following two conditions hold:*

(i) *There is a unique element $x_f \in H$ such that*

$$f(y) = (x_f, y) \quad \forall\, y \in H.$$

(ii) *The norm of $f \in H^*$ is equal to the norm of $x_f \in H$*

$$\|f\|_{H^*} = \|x_f\|_H.$$

It is not difficult to show that Riesz's representation theorem allows us to construct a linear isometry $i : H \to H^*$ defined via

$$i : H \to H^*$$
$$i : x \mapsto f(\cdot) = (x, \cdot)_H.$$

By convention, the Riesz map $\mathcal{R}$ from H^* to H is the function that associates to each linear functional $f \in H^*$ the unique $x_f \in H$ such that

$$f(y) = (x_f, y)_H \qquad \forall\, y \in H.$$

In other words,

$$\mathcal{R} = i^{-1}.$$

This construction has many implications. Some of the more useful implications for applications to mechanics are embodied in the two following theorems.

Theorem 2.6.3. *Every Hilbert space is reflexive.*

Theorem 2.6.4. *The closed unit ball of a Hilbert space is weakly compact.*

This latter theorem follows from Theorem 2.4.3, and the fact that we can show that the weak and weak* topologies coincide on a Hilbert space H.

2.7 Closed Operators on Hilbert Spaces

The simplest class of operators that we will need to consider in this book are those operators that are linear and bounded on normed vector spaces. It is well known, however, and quickly pointed out in a number of introductory texts treating functional analysis, that many of the most frequently encountered operators in applications to mechanics are not bounded linear operators. Perhaps the single largest class of operators of interest to us that are not bounded are differential operators. It is not difficult to construct examples of linear differential operators on normed vector spaces that are not bounded.

To accommodate a larger class of linear operators that includes differential operators, it is conventional to consider closed linear operators. Throughout this section, we will consider a linear operator L that acts from a real Hilbert space V to itself

$$L : \text{dom}(L) \subseteq V \to V.$$

In our original definition of a linear operator acting between normed vector spaces, we did not require that the domain *exhaust* V. We allowed the possibility that there might exist $v \in V$ and $v \notin \text{dom}(L)$. When discussing unbounded linear operators, we will be most interested in those cases for which the $\text{dom}(L) \neq V$! However, we require that the domain of an unbounded linear operator L is a dense subset of V.

Definition 2.7.1. *In a normed space V, a linear operator $L : \text{dom}(L) \subseteq V \to V$ is said to be* densely defined *if $\text{dom}(L)$ is a dense vector subspace of V.*

A densely defined operator $L : \text{dom}(L) \subseteq V \to V$ is said to be *closed* if its graph is closed in the product space $V \times V$, endowed with the usual norm on product spaces.

Definition 2.7.2. *Let $L : \text{dom}(L) \subseteq V \to V$ be a densely defined linear operator on the Hilbert space V. The mapping L is said to be* closed *if its graph $G(L) \subseteq V \times V$ defined as*

$$G(L) \equiv \{\{x, y\} \in V \times V : \ x \in \text{dom}(L), \ y = L(x)\}$$

is closed in $V \times V$.

The following theorem follows immediately from the definition, and frequently is used to characterize closed linear operators instead of the definition.

Theorem 2.7.1. *Let $L : \text{dom}(L) \subseteq V \to V$ be a linear operator. The operator L is closed if and only if the following condition is true: If $\{x_k, Lx_k\}_{k \in \mathbb{N}} \subseteq \text{dom}(L) \times V$ is a sequence such that $x_k \to x^*$ and $Lx_k \to y$, then $x^* \in \text{dom}(L)$ and $y = Lx^*$*

It is a simple exercise to show that every continuous linear operator is closed. We define the *resolvent* set of L to be all $\lambda \in \mathbb{C}$ such that $\lambda I - L$ is bijective (invertible) and the inverse $(\lambda I - L)^{-1}$ is a bounded operator. The *spectrum* of L is the complement of the resolvent in $\mathbb{C}$.

Closed Graph Theorem

Theorem 2.7.2 (Closed Graph Theorem). *Let X and Y be Banach spaces and let $L : \text{dom}(L) \subseteq X \to Y$ be a closed linear operator. If $\text{dom}(L)$ is closed, then L is a bounded linear operator on $\text{dom}(L) \subseteq X$.*

Proof. Since X and Y are Banach spaces, it is not difficult to show that the Cartesian product $X \times Y$ is also a Banach space with norm

$$\|(x, y)\|_{X \times Y} \overset{\triangle}{=} \|x\|_X + \|y\|_Y.$$

By hypothesis, $G(L)$ is closed in $\left(X \times Y, \|(\cdot,\cdot)\| \right)$ and dom(L) is closed in $(X, \|\cdot\|)$. A closed subspace of a complete normed vector space is itself a complete normed vector space. Hence,

$$\left(G(L), \|(\cdot,\cdot)\| \right)$$
$$\left(\mathrm{dom}(L), \|\cdot\|_X \right)$$

are in fact Banach (sub)spaces. Define the mapping

$$i : G(L) \rightarrow \mathrm{dom}(L)$$
$$i : (x, Lx) \mapsto x.$$

Clearly, the mapping i is surjective and injective. The mapping is also linear and bounded since

$$\|i(x, Lx)\|_X = \|x\|_X \leq \|x\|_X + \|Lx\|_Y = \|(x, Lx)\|_{X \times Y}.$$

Hence, i is a bounded, linear operator mapping the Banach space $G(L)$ onto the Banach space dom(L). By the Open Mapping Theorem, the mapping i is an open mapping. That is, if $B \subseteq G(L)$ is an open set, then

$$i(B) \subseteq \mathrm{dom}(L)$$

is an open set. But this is equivalent to stating that the inverse map $i^{-1} :$ dom(L) $\rightarrow G(L)$ is continuous. Since the mapping i is linear and bijective, the mapping i^{-1} is linear. We can conclude that i^{-1} is continuous and linear, and therefore bounded. We can write

$$\|x\|_X + \|Lx\|_Y = \|(x, Lx)\|_{X \times Y} = \|i^{-1}(x)\|_{X \times Y} \leq c\|x\|_X$$

for all $x \in \mathrm{dom}(L)$. But this inequality implies that

$$\|Lx\|_Y \leq c\|x\|_X$$

and L is a bounded linear operator on dom(L).   □

2.8   Adjoint Operators

Let L be a bounded linear operator whose domain is all of H. For each fixed $y \in H$, the mapping

$$x \mapsto (Lx, y)$$

is a bounded, linear operator. Boundedness follows since

$$|(Lx, y)| \leq \|Lx\| \|y\|$$
$$\leq \|L\| \|y\| \|x\|$$
$$= C_y \|x\|.$$

Linearity is also clear

$$x + \alpha z \mapsto \big(L(x + \alpha z), y\big)$$
$$x + \alpha z \mapsto (Lx, y) + \alpha(Lz, y).$$

By the Riesz representation formula, there is a unique $\xi_y \in H$ such that

$$(Lx, y) = (x, \xi_y) \qquad \forall\, x \in H$$

for fixed $y \in H$. However, it is simple to check that the mapping

$$y \mapsto \xi_y$$

so constructed is a bounded, linear operator. By the Riesz representation theorem

$$\|\xi_y\|_H = \sup_{\|x\| \neq 0} \frac{|(Lx, y)|}{\|x\|} \leq \sup_{\|x\| \neq 0} \frac{\|L\|\|x\|\|y\|}{\|x\|} \leq C\|y\|.$$

Definition 2.8.1. *The adjoint operator L^* for a linear bounded operator $L : H \to H$ is defined to be*

$$L^* y = \xi_y$$

for each $y \in H$. In other words, the adjoint operator is the unique mapping L^* such that

$$\big(Lx, y\big) = \big(x, L^* y\big) \qquad \forall\, x,\, y \in H.$$

If L is a densely defined operator $L : \mathrm{dom} L \subseteq H \to H$, we can define the adjoint $L^* : \mathrm{dom}(L^*) \subseteq H \to H$ again via Riesz's representation theorem. For each $y \in H$ the mapping

$$x \mapsto (Lx, y)$$

is a bounded linear operator on the $\mathrm{dom}(L)$. Then there exists a unique $\xi_y \in H$ such that the mapping can be represented as

$$(x, \xi_y) = (Lx, y) \quad \forall\, x \in \mathrm{dom}(L).$$

Note that the uniqueness of ξ_y for $L : H \to H$ is guaranteed by Riesz's representation theorem, while the uniqueness of ξ_y for $L : \mathrm{dom}(L) \subseteq H \to H$ is provided by the requirement for L to be densely defined. Indeed, suppose there exists another ξ_y' such that $(x, \xi_y') = (Lx, y)$ for all $x \in \mathrm{dom}(L)$. Then $(x, \xi_y' - \xi_y) = 0$ for all $x \in \mathrm{dom}(L)$, which implies $\xi_y' - \xi_y = 0$ since $\mathrm{dom}(L)$ is dense.

Definition 2.8.2. *The adjoint operator $L^* : \mathrm{dom}(L^*) \subseteq H \to H$ for a densely defined linear operator $L : \mathrm{dom}(L) \subseteq H \to H$ is defined by*

$$(Lx, y) = (x, L^* y) \qquad \forall\, x \in \mathrm{dom}(L)\ \forall\, y \in \mathrm{dom}(L^*)$$

where

$$\text{dom}(L^*) = \{y \in H : \exists \xi_y \in H \text{ such that } (x, \xi_y) = (Lx, y) \quad \forall x \in \text{dom}(L)\}.$$

The adjoint operator L^* is given by

$$L^* y = \xi_y$$

for such a $\xi_y \in H$, with associated $y \in H$.

Proposition 2.8.1. *Let L be densely defined. Then L^* is a closed operator.*

Proof. Let $\{x_n\}_{n \in \mathbb{N}} \subseteq \text{dom}(L^*)$ be a sequence such that $x_n \to x$ and $L^* x_n \to y$. By definition of adjoint operator, for all $x' \in \text{dom}(L)$ we have $(x', L^* x_n) = (Lx', x_n)$. Hence, taking the limit we obtain $(x', y) = (Lx', x)$. But this means that $x \in \text{dom}(L^*)$ and $L^* x = y$. $\qquad\square$

Definition 2.8.3. *An operator is called symmetric if and only if*

$$(Lx, y) = (x, Ly) \qquad \forall x, y \in \text{dom}(L).$$

In other words, an operator is symmetric if $\text{dom}(L) \subseteq \text{dom}(L^*)$ and $L^* = L$ on $\text{dom}(L)$. For a symmetric operator L, we also have $G(L) \subseteq G(L^*)$.

Definition 2.8.4. *An operator L is self-adjoint if it is symmetric and $\text{dom}(L) = \text{dom}(L^*)$, i.e., $L = L^*$ and $\text{dom}(L) = \text{dom}(L^*)$.*

Obviously, if L is *self-adjoint* then $G(L^*) \subseteq G(L)$.

We may consider a self-adjoint operator as an extension of a corresponding symmetric operator. However, in general, a symmetric operator does not admit a self-adjoint extension. Even if it does then a self-adjoint extension may not be unique. The following proposition establishes a condition for a symmetric operator to be self-adjoint in the case of $\text{dom}(L) = H$.

Proposition 2.8.2. *If L is a bounded symmetric operator and $\text{dom}(L) = H$ then L is self-adjoint.*

We introduce the notions of positive and monotone operators. They play important roles in applications. All operators are assumed to be linear and unbounded.

Definition 2.8.5. *An operator L is positive, if for all $x \in \text{dom}(L)$*

$$(Lx, x) \geq 0.$$

Theorem 2.8.1. *Let L be an operator with $\text{dom}(L)$ in a Hilbert space H. The following are equivalent:*

(i) *The operator L is positive.*

(ii) *For every $\lambda > 0$ the operator $(\lambda I + L)$ is injective, or one-to-one, and*

$$\|(\lambda I + L)^{-1} y\| \leq \frac{1}{\lambda} \|y\|$$

for every $y \in \text{range}(\lambda I + L)$.

Proof. We first show that

(i) → (ii). Let $y \in \text{range}(\lambda I + L)$, so that there exists an $x \in \text{dom}(L)$ such that

$$y = (\lambda I + L)x.$$

We can write

$$((\lambda I + L)x, x) = \lambda(x, x) + (Lx, x) \geq \lambda \|x\|^2 \tag{2.24}$$

which implies that the nullspace of $(\lambda I + L)$ is simply $\{0\}$. (Otherwise, there would exist a $z \in \text{Ker}(\lambda I + L)$ with $z \neq 0$ and

$$0 = ((\lambda I + L)z, z) = \lambda(z, z) + (Lz, z) > 0$$

which is a contradiction.) Thus, $(\lambda I + L)$ is injective. Since $x = (\lambda I + L)^{-1}y$, the inequality (2.24) implies that

$$\begin{aligned} \lambda \|(\lambda I + L)^{-1}y\|^2 &\leq |(\lambda I + L)^{-1}y, y)| \\ &\leq \|(\lambda I + L)^{-1}y\| \|y\| \end{aligned}$$

which implies that

$$\|(\lambda I + L)^{-1}y\| \leq \frac{1}{\lambda}\|y\|.$$

(ii) → (i). For any $x \in \text{dom}(L)$, let $y = (\lambda I + L)x$. We have

$$\begin{aligned} \|x\|^2 &= \|(\lambda I + L)^{-1}y\|^2 \\ &\leq \frac{1}{\lambda^2}\|(\lambda I + L)x\|^2 \\ &\leq \frac{1}{\lambda^2}\left(\lambda^2(x, x) + 2\lambda(Lx, x) + \|Lx\|^2\right). \end{aligned}$$

We conclude that

$$0 \leq 2(Lx, x) + \frac{1}{\lambda}\|Lx\|^2.$$

Since this inequality must hold for all $\lambda > 0$, L is consequently positive. $\square$

Proposition 2.8.3. *Every positive symmetric operator L, densely defined in a Hilbert space H, admits at least one self-adjoint extension, which is also known as a canonical extension or the Friedrichs extension.*

Proof. Since operator L is symmetric and positive, we can define a new inner product by $(x, y)_L = (Lx, y)$ on $\text{dom}(L)$. An extension of $\text{dom}(L)$ by the norm $(x, y)_L$ is a Hilbert space $H_0 \subseteq H$. Obviously, H_0 is a linear subspace of H. Note that by construction, L is self-adjoint on $\text{dom}(L^*) \cap H_0$. Consequently, we obtained a self-adjoint extension L_f of the symmetric positive operator L. The operator L_f is also positive and called the Friedrichs extension of L. The Friedrichs extension is a restriction of operator L^* to the domain $\text{dom}(L^*) \cap H_0$. $\square$

The construction of Friedrichs extensions is very simple. It requires only the knowledge of expression (Lx, y). Because of simplicity of construction, Friedrichs extensions are widely used in the theory of partial differential equations.

Definition 2.8.6. *An operator L is* monotone *if for any $\{x_1, y_1\}$, $\{x_2, y_2\} \in G(L)$*

$$(x_2 - x_1, y_2 - y_1) \geq 0.$$

Proposition 2.8.4. *If L is a symmetric, positive and densely defined operator, then L is monotone.*

Proof. By definition of densely defined operator, $\mathrm{dom}(L)$ is a vector subspace. Hence, if $x_1, x_2 \in \mathrm{dom}(L)$ then $x_1 - x_2 \in \mathrm{dom}(L)$. By definition of positive operator, for any $x_1, x_2 \in \mathrm{dom}(L)$ we have $(L(x_2 - x_1), x_2 - x_1) \geq 0$. However, since L is symmetric, the last inequality is equivalent to $(x_2 - x_1, L(x_2 - x_1)) \geq 0$, which, for the linear operator L, is rewritten as $(x_2 - x_1, Lx_2 - Lx_1) \geq 0$. But this means that L is monotone. $\qquad\square$

A distinguished role will be played by those operators that are, in some sense, the "largest" positive operators. If L and $\tilde{L}$ are linear operators with graphs $G(L)$ and $G(\tilde{L})$, respectively, we say that $(\tilde{L})$ is an extension of L if

$$G(L) \subseteq G(\tilde{L}).$$

An extension $\tilde{L}$ of an operator L is said to be *maximal* if there is no extension of $\tilde{L}$ that contains $G(\tilde{L})$ as a proper subset.

Definition 2.8.7. *An operator L is* maximal monotone *if it is monotone and for any monotone operator L'*

$$G(L') \subseteq G(L).$$

An alternative characterization of a maximal monotone operator is given by the following theorem:

Theorem 2.8.2. *Let L be monotone and $\{x, y\} \in H \times H$. Then L is maximal monotone if*

$$(x - x', y - y') \geq 0 \quad \forall\, \{x', y'\} \in G(L) \quad \Longrightarrow \quad \{x, y\} \in G(L).$$

The following theorem shows that positive self-adjoint operators are maximal monotone operators.

Theorem 2.8.3. *Suppose that L is a symmetric, positive and densely defined operator acting on the Hilbert space H and $\lambda > 0$. The following are equivalent*

(i)   *The operator L is self-adjoint.*
(ii) *For any $\lambda > 0$, the range of $\lambda I + L$ is all of H.*
(iii) *The operator L is maximal monotone.*

Proof. We first show that

(i) $\rightarrow$ (ii). Suppose that L is self-adjoint. We will first show that the orthogonal complement of range($\lambda I + L$) for $\lambda > 0$ is the null vector, and consequently must be dense in H. Suppose that $z \in \{\text{range}(\lambda I + L)\}^{\perp}$. It means that

$$((\lambda I + L)x, z) = 0 \qquad \forall\, x \in \text{dom}(L)$$

Note that $z \in \text{dom}((\lambda I + L)^*)$. Indeed, by definition of adjoint operator, an element z belongs to $\text{dom}((\lambda I + L)^*)$ if there exists $\xi_z \in H$ such that $((\lambda I + L)x, z) = (x, \xi_z)$ for all $x \in \text{dom}(\lambda I + L)$ and $(\lambda I + L)^* z = \xi_z$. In this case, $\xi_z = 0 \in H$ and $(x, 0) \equiv 0$ for all x. Hence, $(\lambda I + L)^* z = 0$, and since L is self-adjoint, we have $Lz = -\lambda z$. But because L is also positive, we conclude

$$(Lz, z) = -\lambda(z, z) \geq 0$$

and, consequently since $\lambda > 0$, we obtain $z = 0$. Thus, the range of $(\lambda I + L)$ is dense in H. We now show that the range of $(\lambda I + L)$ is closed, and therefore equal to all of H. By Theorem 2.8.1, $(\lambda I + L)^{-1}$ is injective and bounded. But since L is self-adjoint, it is closed. Consequently, $(\lambda I + L)$ and $(\lambda I + L)^{-1}$ are also closed. Based on the Closed Graph Theorem, if an operator is closed, bounded and densely defined then its domain is closed, i.e., the domain coincides with the whole space H. Thus, range($\lambda I + L$) = dom($(\lambda I + L)^{-1}$) = H.

(ii) $\rightarrow$ (iii). Since L is symmetric, positive and densely defined, by Proposition 2.8.4, L is monotone. To show that L is maximal monotone, suppose that $\{x_0, y_0\} \in H \times H$ and $(x_0 - x', y_0 - y') \geq 0$ for all $\{x', y'\} \in G(L)$. Because $y' = Lx'$ and $\lambda(x_0 - x', x_0 - x') \geq 0$ for $\lambda > 0$, we have $(x_0 - x', (y_0 + \lambda x_0) - (\lambda I + L)x') \geq 0$. But since range($\lambda I + L$) = H and $\lambda I + L$ is boundedly invertible, there exists $\tilde{x}_\lambda \in \text{dom}(L)$ such that $\tilde{x}_\lambda = (\lambda I + L)^{-1}(y_0 + \lambda x_0)$. Hence, we obtain $(x_0 - x', (\lambda I + L)(\tilde{x}_\lambda - x')) \geq 0$. Because operator $\lambda I + L$ is positive and symmetric, in the dense vector subspace dom(L) we can introduce a new inner product by $(x_1, x_2)^* = (x_1, (\lambda I + L)x_2)$. With this definition, the last inequality is rewritten as $(x_0 - x', \tilde{x}_\lambda - x')^* \geq 0$ for all $x' \in \text{dom}(L)$. But since the dom(L) is dense, this inequality can hold for all $x' \in \text{dom}(L)$ only when $x_0 = \tilde{x}_\lambda$. Thus, $x_0 \in \text{dom}(L)$ and $x_0 = (\lambda I + L)^{-1}(y_0 + \lambda x_0)$. This expression simplifies to $y_0 = Lx_0$. Consequently, $\{x_0, y_0\} \in G(L)$ and, according to Definition 2.8.2, L is maximal monotone.

(iii) $\rightarrow$ (i). Now let L be maximal monotone. Since L is positive and symmetric, by Proposition 2.8.3, there exists a self-adjoint extension L_f of the operator L, and dom(L) $\subseteq$ dom(L_f). Note that by construction of Friedrichs extensions, L_f is positive and symmetric on dom(L_f). Hence, by Proposition 2.8.4, L_f is monotone. However, L is maximal monotone. It means that $G(L_f) \subseteq G(L)$, and hence dom(L_f) $\subseteq$ dom(L). Thus, we conclude that dom(L_f) = dom(L) and, consequently, L is self-adjoint. $\qquad\square$

2.9 Gelfand Triples

The use of Gelfand triples [35], or evolution triples [37], has come to assume a central role in the formulation of elliptic, parabolic and hyperbolic partial differential equations in abstract spaces. In this section, we summarize the most fundamental properties of this construction, and emphasize those attributes of the construction that are most important in characterizing dynamical systems. To begin, we review the notion of the transpose of a linear operator in a Banach space.

Definition 2.9.1. *Let U, V be Banach spaces and let $A \in \mathcal{L}(U, V)$. The transpose $A' \in \mathcal{L}(V^*, U^*)$ is the unique operator defined via the relationship*

$$\langle A'v', u \rangle_{U^* \times U} = \langle v', Au \rangle_{V^* \times V}$$

for all $v' \in V^*$ and $u \in U$.

The following theorem is crucial in determining the structure of the Gelfand triple:

Theorem 2.9.1. *Suppose that U, V are Banach spaces, and that $A \in \mathcal{L}(U, V)$. Then the* range(A) *is dense in V if and only if the dual operator A' is injective.*

Proof. First of all, suppose that the range(A) is dense in V. We know that A' is injective if and only if the kernel of A' is the set $\{0\}$. Let $v' \in \ker(A')$, that is,

$$A'v' = 0 \quad \text{in} \quad U^*.$$

By definition, we can write

$$\begin{aligned} \langle A'v', u \rangle_{U^* \times U} &= 0 \quad \forall u \in U \\ \langle v', Au \rangle_{V^* \times V} &= 0 \quad \forall u \in U. \end{aligned}$$

But this means that v' is "perpendicular" to all elements in a dense subset of V. Hence, $v' = 0$, and A' must be injective.

On the other hand, suppose that A' is injective. In this case, again,

$$\ker(A') = \{v' \in V^* : A'v' = 0\} = \{0\}.$$

Suppose to the contrary that range(A) is not dense in V. Then there is some $z' \in V^*$, $z' \neq 0$ such that

$$\langle z', Au \rangle_{V^* \times V} = 0 \quad \forall u \in U.$$

But in this case,

$$\langle A'z', u \rangle_{U^* \times U} = 0 \quad \forall u \in U.$$

This implies

$$z' \in \ker(A') \quad z' \neq 0$$

which is a contradiction. $\qquad\square$

Definition 2.9.2. *A Banach space U is continuously embedded in V if and only if*

$$\|i(u)\|_V \leq c \|u\|_U$$

for some constant c, where $i : U \to V$ is the embedding map

$$i : U \quad \to \quad V$$
$$i : u \in U \quad \mapsto \quad u \in V.$$

Now suppose we are given a Banach space V and a Hilbert space H such that $V \subseteq H$. Further, suppose that V is dense in H (in the topology on H). We denote by $i : V \to H$ the canonical injection of V into H. That is, the map i associates to each element v in the Banach space V the element v in the Hilbert space H

$$i : v \in V \mapsto v \in H.$$

By applying Definition 2.9.1, we know that the dual mapping i' maps H^* into V^*,

$$i' : H^* \to V^*.$$

Since H is a Hilbert space, the inverse of the Riesz map defines an isometric isomorphism of H onto H^*: for each $x' \in H^*$ there is a $y \in H$ such that

$$\mathcal{R}^{-1} y = x'.$$

Moreover, by considerations detailed in Section 2.6.2, the Riesz map has a representation in terms of the inner product on H

$$x' \in H^* \quad \Longrightarrow \quad \exists y \in H : \quad \langle x', x \rangle_{H^* \times H} = (y, x)_H \quad \forall x \in H.$$

We can summarize the structure of the Gelfand triple as

$$V \overset{i}{\hookrightarrow} H \overset{\mathcal{R}^{-1}}{\to} H^* \overset{i'}{\hookrightarrow} V^*.$$

Usually, we write this construction as

$$V \hookrightarrow H \equiv H^* \hookrightarrow V^*$$

or even as

$$V \hookrightarrow H \hookrightarrow V^*.$$

The Hilbert space H in the above construction is the *pivot space* of the Gelfand triple. The following theorem relates an important attribute of the Gelfand triple that we will use frequently in applications to partial differential equations. The following theorem shows that the duality pairing on $V^* \times V$ can be viewed as the extension by continuity of the inner product $(\cdot, \cdot)_H$ acting on $H \times V$.

Theorem 2.9.2. *Let V be a Banach space continuously and densely embedded in the Hilbert space H.*

(i) *For any $f \in V^*$, there exists a sequence $\{f_n\}_{n \in \mathbb{N}} \subseteq H$ such that*

$$(f_n, v)_H \to \langle f, v \rangle_{V^* \times V} \quad \forall v \in V$$

(ii) *we can write*

$$(f, v)_H = \langle f, v \rangle_{V^* \times V} \quad \forall f \in H, \, v \in V.$$

Proof. This theorem employs an abuse of notation that has become conventional in references to Gelfand triples: all references to the embeddings i, $\mathcal{R}$ and i' are dropped in (i) and (ii) above. These mappings play an important role in the derivation of the theorem. Recall that we have the following structure that defines the Gelfand triple,

$$V \overset{i}{\hookrightarrow} H \overset{\mathcal{R}^{-1}}{\to} H^* \overset{i'}{\hookrightarrow} V^*.$$

From Theorem 2.9.1, $i'(H^*)$ is dense in V^*. Given any $v^* \in V^*$, we can find a sequence

$$\{h_k^*\}_{k \in \mathbb{N}} \subseteq H^*$$

such that

$$i'h_k^* \to v^* \quad \text{in} \quad V^*.$$

For each fixed k, we have

$$\langle i'h_k^*, v \rangle_{V^* \times V} = \langle h_k^*, iv \rangle_{H^* \times H}$$
$$= (\mathcal{R}^{-1}h_k^*, iv)_H.$$

Hence for any arbitrary $v^* \in V^*$, we have found a sequence in H

$$h_k \overset{\triangle}{=} \{\mathcal{R}^{-1}h_k^*\}$$

such that

$$\lim_{k \to \infty} (h_k, iv)_H \to \langle v^*, v \rangle_{V^* \times V}.$$

Thus, (i) is proved. Moreover, each term on the left-hand side of the above limit can, in fact, be construed as an element in V^*. For each $h \in H$, the mapping

$$v \mapsto (iv, h)_H$$

defines a bounded linear functional on V. This follows since $v \mapsto (iv, h)$ is clearly linear and we have

$$|(iv, h)_H| \leq \|iv\|_H \|h\|_H \leq c\|v\|_V \|h\|_H \leq c_h \|v\|_V.$$

It follows that the inner product $(\cdot, \cdot)_H$ can be extended by continuity from $H \times V$ to $V^* \times V$. This extension is given by the duality pairing $\langle \, , \, \rangle_{V^* \times V}$. $\square$

2.10   Bilinear Mappings

The Gelfand triple

$$V \hookrightarrow H \to H^* \hookrightarrow V^*$$

frequently arise in conjunction with a bilinear mapping

$$a(\cdot, \cdot) : V \times V \to \mathbb{R}$$

that satisfies the following assumptions:

Assumption 2.10.1.

(i) *symmetry:*

$$a(x, y) = a(y, x) \qquad \forall\, x, y \in V$$

(ii) *coercivity: There is a constant $\tilde{c} > 0$ such that*

$$a(x, x) \geq \tilde{c}\|x\|_V^2 \qquad \forall\, x \in V$$

(iii) *boundedness: There is a constant $c > 0$ such that*

$$|a(x, y)| \leq c\|x\|_V\|y\|_V \qquad x, y \in V.$$

With this structure, it is possible to consider several operators. Two of the most important are considered briefly. The first operator acts from V to V^*.

Theorem 2.10.1. *Suppose that $a(\cdot, \cdot) : V \times V \to \mathbb{R}$ is bilinear, symmetric, coercive and bounded. Define the operator $\mathcal{A} : V \to V^*$ by*

$$\langle \mathcal{A}u, v \rangle_{V^* \times V} \overset{\triangle}{=} a(u, v) \qquad u, v \in V.$$

Then $\mathcal{A}$ is

(i) *linear,*
(ii) *bounded, and*
(iii) *boundedly invertible (i.e., $\mathcal{A}^{-1}$ exists and is bounded).*

Proof. The proof that $\mathcal{A}$ is linear is straightforward. We have

$$\langle \mathcal{A}(\alpha u + \beta w), v \rangle_{V^* \times V} = a(\alpha u + \beta w, v)$$
$$= \alpha a(u, v) + \beta a(w, v)$$
$$= \alpha \langle \mathcal{A}u, v \rangle_{V^* \times V} + \beta \langle \mathcal{A}w, v \rangle_{V^* \times V} \qquad \forall\, v \in V$$

so that

$$\mathcal{A}(\alpha u + \beta w) = \alpha \mathcal{A}u + \beta \mathcal{A}w$$

for all $u, w \in V$. By the boundedness of the bilinear form $a(\cdot, \cdot)$, we can write

$$|\langle \mathcal{A}u, v \rangle_{V^* \times V}| = |a(u, v)| \leq c\|u\|_V\|v\|_V.$$

Thus, for each $u \in V$

$$\|\mathcal{A}u\|_{V^*} = \sup_{v \neq 0} \frac{|\langle \mathcal{A}u, v \rangle_{V^* \times V}|}{\|v\|} \leq c\|u\|_V.$$

Therefore, we have

$$\|\mathcal{A}\| = \sup_{u \in V} \frac{\|\mathcal{A}u\|_{V^*}}{\|u\|_V} \leq c.$$

By the coercivity of $\mathcal{A}$, we know that the kernel of $\mathcal{A}$ is $\{0\}$, and $\mathcal{A}$ is invertible. Suppose $v^* = \mathcal{A}v$, then

$$\tilde{c}\|v\|_V^2 \leq a(v, v) = |\langle \mathcal{A}v, v \rangle_{V^* \times V}| \leq \|\mathcal{A}v\|_{V^*}\|v\|_V$$

or

$$\|\mathcal{A}^{-1}v^*\|_V \leq \frac{1}{\tilde{c}}\|v^*\|_{V^*}.$$

This is simply a statement that $\mathcal{A}^{-1}$ is uniformly bounded,

$$\|\mathcal{A}^{-1}\| < \frac{1}{\tilde{c}}. \qquad \square$$

Thus, we see that the bilinear form $a(\cdot, \cdot)$ induces a bounded, linear operator, and this operator has a bounded inverse. It is also possible to define an unbounded, but closed operator $\mathbb{A}$ associated with the bilinear form $a(\cdot, \cdot)$. We define the domain of $\mathbb{A}$ as

$$\begin{aligned} \text{dom}(\mathbb{A}) &= \{v \in V \subset H : \mathcal{A}v \in H\} \\ &= \mathcal{A}^{-1}(H). \end{aligned} \qquad (2.25)$$

The operator $\mathbb{A}$ is then defined to be the restriction of $\mathcal{A}$ to $\text{dom}(\mathbb{A})$

$$\mathbb{A} = \mathcal{A}|_{\text{dom}(\mathbb{A})}. \qquad (2.26)$$

Some of the most commonly employed properties of $\mathbb{A}$ are summarized in Theorem 2.10.2.

Theorem 2.10.2. *Suppose that $a(\cdot, \cdot) : V \times V \to \mathbb{R}$ is bilinear, symmetric, coercive and bounded. Define the operator $\mathbb{A}$ as in Equations 2.25 and 2.26. Then*

(i) $\mathbb{A}$ *is linear,*
(ii) $\text{dom}(\mathbb{A})$ *is dense in H and*
(iii) $\mathbb{A}$ *is closed.*

Proof. The linearity of the operator $\mathbb{A}$ follows immediately from the linearity of the operator $\mathcal{A}$. To show that the domain of $\mathbb{A}$ is dense in H, let

$$w \in \text{dom}(\mathbb{A})^{\perp} \subset H.$$

If this is the case, we have

$$(w, v)_H = 0 \qquad \forall v \in \text{dom}(\mathbb{A}).$$

Since $\mathbb{A} = \mathcal{A}|_{\text{dom}}(\mathbb{A})$ and $\mathcal{A}$ is proved to be invertible, $\mathbb{A}$ is invertible on $\text{dom}(\mathbb{A})$. Consequently, there is a $u \in \text{dom}(\mathbb{A})$ such that

$$\mathbb{A}u = w.$$

Thus, we have

$$(\mathbb{A}u, v)_H = 0 \qquad \forall v \in \text{dom}(\mathbb{A})$$

and because $\mathbb{A}$ is symmetric,

$$(\mathbb{A}v, u)_H = 0 \qquad \forall v \in \text{dom}(\mathbb{A}).$$

By the definition of $\mathbb{A}$,

$$0 = (\mathbb{A}v, u)_H = a(v, u) = \langle \mathcal{A}v, u \rangle_{V^* \times V}.$$

Finally, recall that

$$\langle \mathcal{A}v, u \rangle_{V^* \times V} = (\mathcal{A}v, u)_H.$$

In other words, the duality pairing $\langle \cdot, \cdot \rangle_{V^* \times V}$ can be interpreted as the extension of $(\cdot, \cdot)_H$ on $H \times V$ to $V^* \times V$. Combining Equations 2.25 and 2.26, we see that

$$(\mathcal{A}v, u)_H = 0 \qquad \forall v \in \text{dom}(\mathbb{A})$$

since $\mathcal{A}$ maps $\text{dom}(\mathbb{A})$ onto H, it follows that $u = 0$ and that $w = \mathbb{A}u = 0$. We have shown that $\text{dom}(\mathbb{A})^\perp = \{0\}$ and, consequently $\text{dom}(\mathbb{A})$ is dense in H. Finally, we show that $\mathbb{A}$ is closed. We must show that if $\{u_k\}_{k \in \mathbb{N}} \subseteq \text{dom}(\mathbb{A})$ and

$$u_k \to u_0 \quad \text{in} \quad H \quad \text{and} \quad \mathbb{A}u_k \to \xi \quad \text{in} \quad H,$$

then

$$u_0 \in \text{dom}(\mathbb{A}) \quad \text{and} \quad \mathbb{A}u_0 = \xi.$$

Define

$$v_k \stackrel{\triangle}{=} \mathbb{A}u_k \in H$$

$$u^* \stackrel{\triangle}{=} \mathcal{A}^{-1}\xi.$$

By the triangle inequality, we have

$$\|u_0 - u^*\|_H \le \|u_0 - u_k\|_H + \|u_k - u^*\|_H.$$

By the continuity of the embedding

$$V \hookrightarrow H$$

we have the bound

$$\begin{aligned}
\|u_0 - u^*\|_H &\leq \|u_0 - u_k\|_H + c'\|u_k - u^*\|_V \\
&= \|u_0 - u_k\|_H + c'\|\mathcal{A}^{-1}v_k - \mathcal{A}^{-1}\xi\|_V.
\end{aligned}$$

By Theorem 2.10.1, the operator $\mathcal{A}^{-1}$ is bounded from V^* to V, i.e., $\|\mathcal{A}^{-1}\| \leq \frac{1}{\tilde{c}}$. Consequently, we obtain

$$\|u_0 - u^*\|_H \leq \|u_0 - u_k\|_H + \frac{c'}{\tilde{c}}\|v_k - \xi\|_{V^*}. \qquad (2.27)$$

By hypothesis, we know that

$$v_k \rightarrow \xi \quad \text{in} \quad H.$$

However, the continuous embedding

$$H \hookrightarrow V^*$$

implies that

$$\|v_k - \xi\|_{V^*} \leq c''\|v_k - \xi\|_H$$

so that inequality (2.27) becomes

$$\|u_0 - u^*\|_H \leq \|u_0 - u_k\|_H + \frac{c'c''}{\tilde{c}}\|v_k - \xi\|_H.$$

Thus we have

$$u_0 = u^* = \mathcal{A}^{-1}\xi. \qquad \square$$

Chapter 3

Common Function Spaces
in Applications

Throughout this book, we have presented various abstract spaces whenever they could illustrate a particular topic in functional analysis. For example, we have discussed

- l_p in Examples 1.3.7, 1.5.4
- $C[a, b]$ in Examples 1.3.6, 1.5.3, Theorem 2.6.1
- $BV[a, b]$ in Example 1.5.5

Each of these spaces plays a central role in a host of applications. In this chapter, we will present some of the abstract spaces that are essential to applications in control and mechanics. We include a discussion of the

- $L^p(\Omega)$, Lebesgue spaces of integrable functions
- $W^{p,q}(\Omega)$, $H^p(\Omega)$, Sobolev spaces of weakly differentiable functions
- $L^p((0, T), X)$, Bochner spaces of integrable functions
- $W^{p,q}((0, T), X, Y)$, Bochner spaces of weakly differentiable functions

3.1 The L^p Spaces

If their frequency of use is a measure of indispensability, the spaces of Lebesgue integrable functions may be the single most important class of functions having practical use to engineers and scientists.

Definition 3.1.1. *Let $1 \leq p < \infty$ and suppose Ω is a Lebesgue measurable subset of $\mathbb{R}^n$. We define*

$$L^p(\Omega) \triangleq \left\{ f : \Omega \to \mathbb{R}, \quad \|f\|_{L^p(\Omega)} \triangleq \left(\int_\Omega |f|^p dx \right)^{1/p} < \infty \right\}$$

while for $p = \infty$, we define

$$L^\infty(\Omega) \triangleq \left\{ f : \Omega \to \mathbb{R}, \quad \|f\|_{L^\infty(\Omega)} \triangleq \operatorname*{ess\,sup}_{x \in \Omega} |f(x)| < \infty \right\}.$$

The L^p spaces are utilized in nearly every example in this book. In other examples, vector-space-valued generalizations of the L^p spaces are employed. We recall several properties of these spaces that are important in applications.

Theorem 3.1.1. *For $1 \leq p \leq \infty$, $L^p(\Omega)$ is a Banach space. The spaces $L^p(\Omega)$ are reflexive if $1 < p < \infty$ and separable for $1 \leq p < \infty$. The topological dual $\left(L^p(\Omega)\right)^*$ is isometrically isomorphic to $L^q(\Omega)$ for $1 < p < \infty$, where*

$$\frac{1}{p} + \frac{1}{q} = 1.$$

Any bounded linear functional $\phi_g(\cdot)$ defined on $L^p(\Omega)$ is realized by the integral representation

$$f \mapsto \phi_g(f) = \int_\Omega f g \, dx, \qquad f \in L^p(\Omega), \quad g \in L^q(\Omega).$$

Compactness properties of the Lebesgue spaces $L^p(\Omega)$ are straightforward if $1 < p < \infty$, and more complicated in the case $p = \infty$ or $p = 1$. The reader may consult [23] for details. Since $L^p(\Omega)$ is a reflexive Banach space for $1 < p < \infty$, bounded sets are weakly sequentially compact. That is, if a sequence $\{f_k\}_{k\in\mathbb{Z}}$ is bounded in $L^p(\Omega)$

$$\|f_k\|_{L^p(\Omega)} \leq C \leq \infty$$

then the sequence contains a weakly convergent subsequence. It is important to note that $L^\infty(\Omega) = \left(L^1(\Omega)\right)^*$, but the topological dual of $L^\infty(\Omega)$ is not $L^1(\Omega)$. In fact, $\left(L^\infty(\Omega)\right)^*$ can be characterized in terms of a collection of additive set functions, [26], but we will not have the occasion to use this space in this book. Figure 3.1 shows how to define a weak* topology on $L^\infty(\Omega)$.

According to Figure 3.1, a weak* topology on $L^\infty(\Omega)$ is the weak topology generated by image $\mathcal{I}\left(L^1(\Omega)\right)$ in $\left(L^1(\Omega)\right)^{**}$ of the canonical injection $\mathcal{I}$. If $\{f_k\}_{k\in\mathbb{N}} \subseteq L^\infty(\Omega)$, then this sequence converges to f in the weak* topology on $L^\infty(\Omega)$ if

$$\int_\Omega g f_k \, dx \to \int_\Omega g f \, dx$$

for all $g \in L^1(\Omega)$. Since $L^1(\Omega)$ is separable, bounded sequences in $L^\infty(\Omega) \equiv \left(L^1(\Omega)\right)^*$ are sequentially weakly* compact. That is, if $\{f_k\}_{k\in\mathbb{N}} \in L^\infty(\Omega)$ and

$$\|f_k\|_\infty < C$$

there is a subsequence $\{f_{k_j}\}_{k_j\in\mathbb{N}}$ such that

$$\int_\Omega f_{k_j} g \, dx \to \int_\Omega f g \, dx$$

for all $g \in L^1(\Omega)$. The weak relative compactness of $L^1(\Omega)$, in comparison, requires additional analysis. Specifically, we need to define equi-integrability.

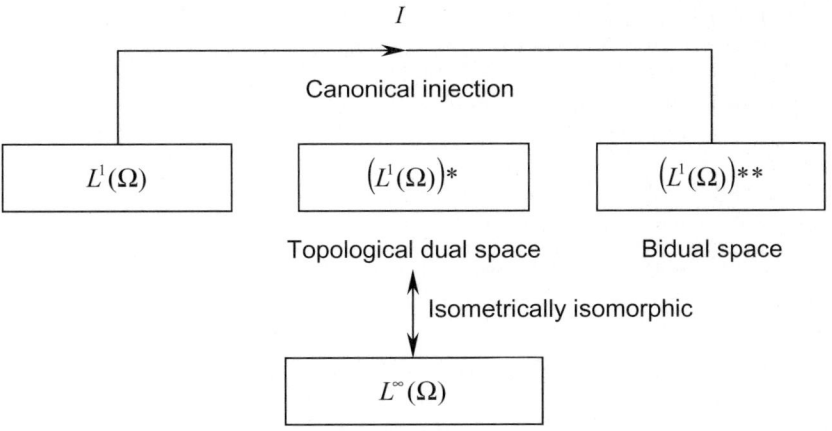

Figure 3.1: $L^1(\Omega)$, its Dual and Bidual

Definition 3.1.2. *A collection of functions $\{f_k\}_{k\in\mathbb{Z}}$ is said to be equi-integrable if for each $\epsilon > 0$ there is a $\delta > 0$ such that for any Lebesgue measurable set E with $|E| < \delta$, we have*

$$\int_E |f_k| dx < \epsilon$$

uniformly in $k \in \mathbb{Z}$.

The following theorem provides useful characterizations of compactness in $L^1(\Omega)$.

Theorem 3.1.2. *A bounded set in $L^1(\Omega)$ is weakly relatively compact if and only if it is equi-integrable.*

3.2 Sobolev Spaces

In discussing classical examples of Banach spaces, we introduced the set

$$C^m(\overline{\Omega}) \equiv \left\{ f : f \text{ is } m \text{ times (classically) continuously differentiable on } \overline{\Omega} \subseteq \mathbb{R}^n \right\}$$

which proved to be a Banach space when endowed with the norm

$$\|f\|_{C^m(\overline{\Omega})} = \sum_{|k| \leq m} \sup_{x \in \overline{\Omega}} \left| D^k f(x) \right|.$$

It is important to note that the derivative used to define $C^m(\overline{\Omega})$ is the classical derivative, first studied in an introductory course in differential calculus. For several reasons, abstract spaces based on the existence of only classical derivatives have not proven to provide a sufficiently rich foundation for the study of partial differential equations. The theory of Sobolev spaces has been developed by generalizing the notion of a classical derivative and introducing the idea of weak or generalized derivatives.

3.2.1   Distributional Derivatives

By a multi-index over $(\mathbb{Z}_0^+)^n$ we will mean an n-tuple $\alpha = (\alpha_1, \alpha_2, \ldots, \alpha_n)$ of nonnegative integers. The length of the multi-index, denoted by $|\alpha|$ is simply

$$|\alpha| = \sum_{i=1}^{n} \alpha_i.$$

Multi-indices are effective for representing the partial derivatives of real-valued functions defined over open subsets Ω of $\mathbb{R}^n$. We write

$$
\begin{aligned}
D^\alpha f &= \frac{\partial^{\alpha_1}}{\partial x_1^{\alpha_1}} \frac{\partial^{\alpha_2}}{\partial x_2^{\alpha_2}} \cdots \frac{\partial^{\alpha_n}}{\partial x_n^{\alpha_n}} f \\
&= \frac{\partial^{\alpha_1 + \alpha_2 + \cdots + \alpha_n}}{\partial x_1^{\alpha_1} \partial x_2^{\alpha_2} \cdots \partial x_n^{\alpha_n}} f
\end{aligned}
$$

where $f : \mathbb{R}^n \to \mathbb{R}^n$ and $x = (x_1, x_2, \ldots, x_n) \in \mathbb{R}^n$. Heuristically speaking, Sobolev spaces are constructed by replacing the rather strict notion of a classical derivative by a weak derivative defined via the integration by parts formula. Recall that the integration by parts formula of classical differential calculus can be written formally as

$$\int_\Omega (D^\alpha f) \cdot \phi \, dx = (-1)^{|\alpha|} \int_\Omega f \cdot D^\alpha \phi \, dx + \underbrace{\text{``boundary terms''}}_{\to 0}.$$

We are interested in considering cases in which the boundary terms above vanish. Of course, there are a variety of conditions under which the above expression makes sense, classically speaking. Let us define $C_0^\infty(\mathbb{R}^n)$ or $\mathcal{D}(\mathbb{R}^n)$ to be the set of all smooth (differentiable of any order in the classical sense) functions with support contained in a compact subset of $\mathbb{R}^n$. By the support of any function f, we mean the closure of the set on which f is nonzero. Because we will have occasion to use this space of functions frequently, we refer to it as the space of test functions. Similarly, $C_0^\infty(\Omega)$ or $\mathcal{D}(\Omega)$ will be used to denote the set of all smooth functions with compact support contained in the open set Ω. If the underlying set Ω or $\mathbb{R}^n$ does not play a role in a particular definition, we simply refer to the set of test functions as $\mathcal{D}$.

The actual mechanics of defining rigorously the weak derivative of functions from the integration by parts formula above is a lengthy process, one that has been studied extensively elsewhere. For completeness, we will briefly outline the development. The interested reader is referred to any of the numerous excellent texts on the subject for details [1, 35].

The collection of infinitely differentiable functions with compact support, $\mathcal{D}(\Omega)$, can be made a topological space if we define convergence of functions in $\mathcal{D}(\Omega)$ as follows:

Definition 3.2.1. *A sequence of functions* $\{f_k\}_{k\in\mathbb{N}} \subseteq \mathcal{D}(\Omega)$ *converges in the sense of test functions to* $\phi \in \mathcal{D}(\Omega)$ *if and only if the following two conditions hold:*

(i) *there exists a single compact set* $\Theta \subseteq \Omega$ *such that*

$$\text{support}(f_k) \subseteq \Theta \quad \forall\, k \in \mathbb{N}$$

(ii) *For any multi-index* $\alpha \in (\mathbb{Z}_0^+)^n$,

$$D^\alpha f_k(x) \to D^\alpha \phi(x)$$

uniformly in Θ.

In fact the topology on $\mathcal{D}(\Omega)$ induced by defining convergence of elements as in Definition 1.2.14 is the strict inductive limit topology (see, e.g., Wilansky in [34]). Thus, $\mathcal{D}$ is a vector space, and more generally, a topological vector space. It makes sense, consequently, to consider the collection of all continuous linear functionals on $\mathcal{D}(\Omega)$. We represent the topological dual of the test functions $\mathcal{D}(\Omega)$ by $\mathcal{D}^*$. That is, we define

$$\mathcal{D}^* = \big\{ f : \mathcal{D} \to \mathbb{R} : \quad f(\cdot) \text{ is linear on } \mathcal{D} \text{ and} \\ \phi_n \to \phi \text{ in } \mathcal{D} \implies f(\phi_n) \to f(\phi) \in \mathbb{R} \big\}.$$

Following the notational convention introduced in Section 2.2, we denote the action of any $f \in \mathcal{D}^*$ on $\phi \in \mathcal{D}$ by the familiar duality pairing notation,

$$\langle f, \phi \rangle_{\mathcal{D}^* \times \mathcal{D}} = f(\phi) \qquad \forall\, \phi \in \mathcal{D},\ f \in \mathcal{D}^*.$$

The reader should note, however, that the space of test functions $\mathcal{D}$ is not metrizable, and not a normed vector space. With this structure in place, we can now define the distributional derivative, that is the derivative of a continuous linear functional on the space of test functions.

Definition 3.2.2. *Let* $T \in \mathcal{D}^*$. *We define the derivative of the distribution* T *of order* m *to be the unique continuous linear functional* $T^{(m)} \in \mathcal{D}^*$ *satisfying*

$$\langle T^{(m)}, \phi \rangle_{\mathcal{D}^* \times \mathcal{D}} = (-1)^m \langle T, \phi^{(m)} \rangle_{\mathcal{D}^* \times \mathcal{D}} \tag{3.1}$$

where $\phi^{(m)} = \frac{\partial^m \phi}{\partial x^m}$ *denotes the usual classical derivative of the test function* $\phi \in \mathcal{D}$.

Several comments are in order to justify Equation (3.1) in the above definition. First, by the definition of $\mathcal{D}^*$, if $T \in \mathcal{D}^*$ and $\phi \in \mathcal{D}$, the right-hand side is well defined, since $\phi^{(m)} \in \mathcal{D}$. In addition, the map

$$\phi \mapsto \langle T, \phi^{(m)} \rangle_{\mathcal{D}^* \times \mathcal{D}}$$

is clearly linear for each $\phi \in \mathcal{D}$. Finally, suppose that ϕ_n converges to ϕ in the sense of test functions $\mathcal{D}$. We want to show that the map $\phi \mapsto \langle T, \phi^{(m)} \rangle_{\mathcal{D}^* \times \mathcal{D}}$ is continuous. We must show that

$$\langle T, \phi_n^{(m)} \rangle_{\mathcal{D}^* \times \mathcal{D}} \to \langle T, \phi^{(m)} \rangle_{\mathcal{D}^* \times \mathcal{D}}.$$

But this follows immediately from the definition of T as an element of $\mathcal{D}^*$, and the fact that

$$\psi_n \to \psi \in \mathcal{D}$$

where $\psi_n \equiv \phi_n^{(m)}$ and $\psi \equiv \phi^{(m)}$.

Of course, this definition was not chosen at random; rather it agrees with our usual notion of differentiability when T corresponds to a classically differentiable function. The definition also extends the classical notion of differentiability in several other respects.

Theorem 3.2.1. *Let T be a distribution in the topological dual $\mathcal{D}^*$ of the space of test functions $\mathcal{D}$.*

(i) *The distributional derivative of T of order α exists for any multi-index α, and can be written as*

$$\langle T^{(\alpha)}, \phi \rangle_{\mathcal{D}^* \times \mathcal{D}} = (-1)^{|\alpha|} \langle T, \phi^{(\alpha)} \rangle_{\mathcal{D}^* \times \mathcal{D}}.$$

(ii) *Every $f \in C_0^m(\Omega)$ defines a distribution F via the formula*

$$\langle F, \phi \rangle_{\mathcal{D}^* \times \mathcal{D}} = \int_\Omega f\phi \, dx \quad \forall \phi \in \mathcal{D}.$$

We can identify the distributional derivative of F with the classical derivative of f via the formula

$$\langle F^{(m)}, \phi \rangle_{\mathcal{D}^* \times \mathcal{D}} = (-1)^m \langle F, \phi^{(m)} \rangle_{\mathcal{D}^* \times \mathcal{D}}$$
$$= (-1)^m \int_\Omega f\phi^{(m)} dx = \int_\Omega f^{(m)}\phi \, dx.$$

At this point, we should note that several classes of functions can be associated, or identified with distributions. For example, for any function $f \in L^2$, the Lebesgue integral

$$\int_\Omega f\phi \, dx$$

is well defined for any $\phi \in \mathcal{D}$. The integral defines a functional that is linear in the test function ϕ, and continuous in the sense of convergence of test functions. That is, if $\phi_n \to \phi$ in $\mathcal{D}$, then

$$\left| \int_\Omega f(\phi_n - \phi)dx \right| \le c \sup_{x \in \Omega} |\phi_n(x) - \phi(x)| \, \|f\|_{L^2}, \quad c = \left(\int_\Omega dx \right)^{1/2}$$

which converges to zero as n approaches infinity. We can consequently define a distribution T_f via the formula

$$\langle T_f, \phi \rangle_{\mathcal{D}^* \times \mathcal{D}} \equiv \int_\Omega f\phi \, dx.$$

We also know that $\mathcal{D} \subseteq L^2$, and we can view L^2 as being intermediate between $\mathcal{D}$ and $\mathcal{D}^*$ in the sense that

$$\mathcal{D} \subseteq L^2 \subseteq \mathcal{D}^*.$$

3.2.2 Sobolev Spaces, Integer Order

With the definition of distributions, or generalized functions, and their derivatives, we can now define the Sobolev spaces of integer order.

Definition 3.2.3. *Let Ω be an open and bounded domain in $\mathbb{R}^n$, let $1 \leq p \leq \infty$ and let $k > 0$ be an integer. The Sobolev space $W^{k,p}(\Omega)$ consists of the set of all functions contained in $L^p(\Omega)$ whose distributional derivatives through order k are contained in $L^p(\Omega)$. That is,*

$$W^{k,p}(\Omega) = \left\{ f \in L^p(\Omega) : \|D^\alpha f\|_{L^p(\Omega)} < \infty \quad \forall |\alpha| \leq k \right\}.$$

We have already discussed how $L^p(\Omega)$ can be interpreted as a subset of $\mathcal{D}^*$. It should be clear that when we speak of the distributional derivative of a function $f \in L^p(\Omega)$, we mean the distributional derivative of the distribution T_f generated by f. With this interpretation in mind, it is conventional to simplify notation and dispense with the qualifier "... distributional derivative of the distribution generated by f." It is crucial to the study of partial differential equations that the Sobolev spaces are in fact Banach spaces.

Theorem 3.2.2. *Let Ω be an open and bounded domain in $\mathbb{R}^n$, let $1 \leq p \leq \infty$ and let $k > 0$ be an integer. The Sobolev space $W^{k,p}(\Omega)$ is a Banach space when endowed with the norm*

$$\|f\|_{W^{k,p}(\Omega)}^p \equiv \sum_{|\alpha| \leq k} \|D^\alpha f\|_{L^p(\Omega)}^p.$$

The Sobolev space $W^{k,p}(\Omega)$ is reflexive for $1 < p < \infty$ and separable for $1 \leq p < \infty$.

Proof. We only sketch the proof of this well-known result, and the reader is referred to [35] for a detailed presentation. Suppose that $\{f_i\}_{i \in \mathbb{N}}$ is Cauchy in $W^{k,p}(\Omega)$. We have

$$\|f_i - f_j\|_{W^{k,p}(\Omega)}^p \to 0 \quad \text{as} \quad i, j \to \infty. \tag{3.2}$$

For each fixed α, with $|\alpha| \leq k$, we must have

$$D^\alpha f_i \to \xi^\alpha$$

in $L^p(\Omega)$. This fact follows since $L^p(\Omega)$ is complete, and Equation (3.2) implies that

$$\|D^\alpha f_i - D^\alpha f_j\|_{L^p(\Omega)} \to 0$$

for each $|\alpha| \leq k$. Of course, for $\alpha = 0$, we obtain

$$f_i \to f \in L^p(\Omega).$$

From the definition of a distributional derivative we conclude that

$$\xi^\alpha = D^\alpha f$$

since

$$\langle D^\alpha f_i, \phi \rangle = \int_\Omega (D^\alpha f_i) \cdot \phi \, dx = (-1)^{|\alpha|} \int_\Omega f_i D^\alpha \phi \, dx.$$

But it is straightforward to show that the left most term converges to $\langle \xi^\alpha, \phi \rangle_{\mathcal{D}^* \times \mathcal{D}}$, while the right most term converges to $(-1)^{|\alpha|} \int_\Omega f D^\alpha \phi \, dx$.   □

For $1 \leq p < \infty$, the Sobolev spaces $W^{k,p}(\Omega)$ are separable, and if $1 < p < \infty$ they are reflexive. Of particular interest in applications is the case when $p = 2$.

Definition 3.2.4. *The Sobolev space $W^{k,2}(\Omega)$ is denoted by $H^k(\Omega)$.*

We then find that $W^{k,2}(\Omega)$ is denoted by $H^k(\Omega)$ because it is a Hilbert space.

Theorem 3.2.3. *The Sobolev space $H^k(\Omega) \triangleq W^{k,2}(\Omega)$ is a Hilbert space with the inner product*

$$(f,g)_{H^k(\Omega)} = \sum_{|\alpha| \leq k} \left(D^\alpha f, D^\alpha g \right)_{L^2(\Omega)}.$$

3.2.3 Sobolev Spaces, Fractional Order

The Sobolev spaces of integer order have an intuitive interpretation. For some function $f \in W^{k,p}(\Omega)$, the integer order k gives the number of weak derivatives that exist in the space $L^p(\Omega)$. However, many applications require extension of the definition of Sobolev spaces to include fractional order spaces. The definition of fractional order Sobolev spaces can be accomplished by at least three distinct techniques:

(1) Fourier transform methods,
(2) Interpolation space methods, and
(3) Introduction of the Sobolev-Slobodeckij Norm.

The Fourier transform methods and interpolation space methods are instructive and provide interesting interpretations of the fractional order Sobolev spaces. However, their discussion and explanation are simply too lengthy for this brief overview. The interested reader is referred to [2, 29, 31] for detailed discussions of each method. The clear advantage of the third approach above is that it is succinct. Unfortunately, the introduction of the Sobolev-Slobodeckij norm simply does not give much insight into the nature of these spaces. We will rely on the discussions of domains, traces and duals of fractional order Sobolev spaces in succeeding sections to provide insight into the nature of these spaces.

Definition 3.2.5. *Let $\Omega \subseteq \mathbb{R}^n$ be open and bounded. Let k be a nonnegative real number having the form*

$$k = [k] + q \geq 0$$

where $[k]$ is the integer part of k, $[k] \geq 0$ and $q \in (0,1)$. We define $W^{k,p}(\Omega)$ to be the subspace of $W^{[k],p}(\Omega)$ with finite Sobolev-Slobodečkij norm

$$\|f\|_{W^{k,p}(\Omega)} \overset{\triangle}{=} \left\{ \|f\|^p_{W^{[k],p}(\Omega)} + \sum_{|m|=[k]} \int_\Omega \int_\Omega \frac{|D^m f(x) - D^m f(y)|^p}{|x-y|^{n+qp}} \, dxdy \right\}^{\frac{1}{p}}$$

where $1 < p < \infty$. If $p = \infty$, $W^{k,\infty}(\Omega)$ is defined to be the subspace of $W^{[k],\infty}(\Omega)$ for which the norm

$$\|f\|_{W^{k,\infty}(\Omega)} \overset{\triangle}{=} \|f\|_{W^{[k],\infty}(\Omega)} + \max_{|m|=[k]} \operatorname*{ess\,sup}_{\substack{x,y \in \Omega \\ x \neq y}} \frac{|D^m f(x) - D^m f(y)|}{|x-y|^q}$$

is finite.

The Sobolev spaces are among the most common function spaces in use in diverse fields of mechanics. Among the numerous important attributes of these spaces, the embedding, density and compactness properties are indispensable in applications. The theorems in the remainder of this section, which describe the compactness, embedding, density and boundary behavior of functions in Sobolev spaces require specific notions of the regularity of the underlying domain Ω. The precise statement of the regularity of a given domain Ω can be quite abstract, depending on the degree of generality [35]. Fortunately, in our applications, some fairly intuitive notions of regularity will suffice. We will treat domains in this text that are either C^m smooth domains, or in some instances, Lipschitz domains. In either case, the regularity of the domain is intrinsically tied to the smoothness of the curves (or parameterizations) of the boundary $\partial\Omega$ of the domain Ω.

Definition 3.2.6. *Suppose that Ω_1, $\Omega_2 \subseteq \mathbb{R}^n$ and $f : \Omega_1 \to \Omega_2$. The mapping f is said to be C^m-diffeomorphic between Ω_1 and Ω_2 if:*

(1) *f is homeomorphic from $\bar{\Omega}_1$ to $\bar{\Omega}_2$ (f is continuous from $\bar{\Omega}_1$ to $\bar{\Omega}_2$ and has continuous inverse f^{-1} from $\bar{\Omega}_2$ to $\bar{\Omega}_1$).*

(2) *f and f^{-1} are classically differentiable up to order m on $\bar{\Omega}_1$ and $\bar{\Omega}_2$, respectively. Specifically, we have $f \in C^r(\bar{\Omega}_1)$, $f^{-1} \in C^r(\bar{\Omega}_2)$ for $1 \leq r \leq m$.*

(3) *there exist two constants $\underline{c}$ and $\bar{c}$ such that the Jacobian $|Df|$ of the transformation satisfies*

$$\underline{c} < |Df(x)| < \bar{c} \qquad \forall\, x \in \Omega$$

We can now define a C^m-smooth domain:

Definition 3.2.7. *Let Ω be a domain in $\mathbb{R}^n$. We say that Ω is of class C^m if for every $x \in \partial\Omega$, we can find an open set O_x containing x and a coordinate chart ϕ_x such that:*

(1) *ϕ_x is C^m-diffeomorphic from O_x to $B_1(0) \subseteq \mathbb{R}^n$, the unit ball in $\mathbb{R}^n$.*

Moreover, for some $k \in 1, 2, \ldots, n$ we have

(2) $\phi_x(O_x \cap \partial\Omega) = \{x \in B_1(0) : x_k = 0\}$
(3) $\phi_x(O_x \cap \Omega) = \{x \in B_1(0) : x_k > 0\}$
(4) $\phi_x(O_x \cap (\mathbb{R}^n / \Omega)) = \{x \in B_1(0) : x_k < 0\}$

Figure 3.2 shows the schematic representation for ϕ_x diffeomorphic from O_x to $B_1(0)$.

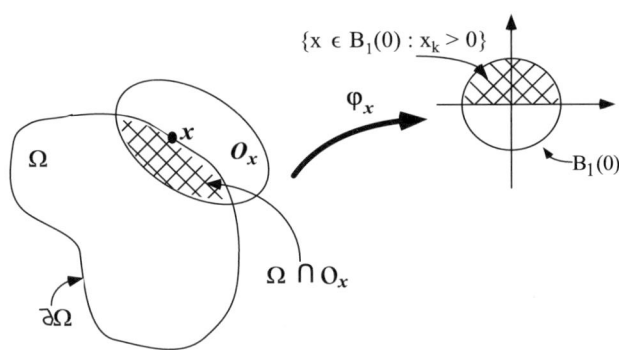

Figure 3.2: ϕ_x Diffeomorphic from O_x to $B_1(0)$

A C^m domain can be a rather smooth domain. To allow a finer classification of domains, it is possible to state Definition 3.2.7 in terms of $C^{k,\alpha}$ Hölderian coordinate charts. The resulting domains are denoted (k, α)-Hölderian domains, or are said to be of class $C^{k,\alpha}$.

Recall that a space of real-valued functions $C^{k,\alpha}$, or (k, α)-Hölderian functions were defined in Example 1.5.8. The special case where $(k, \alpha) = (0, 1)$ defines Lipschitz continuous functions. A straightforward extensions of these examples will enable us to define class $C^{k,\alpha}$ domains.

Definition 3.2.8. *Let X and Y be normed vector spaces. A mapping $f : X \to Y$ is called Lipschitz if for all $x_1, x_2 \in X$*

$$\|f(x_2) - f(x_1)\|_Y \leq C \|x_2 - x_1\|_X$$

where $C \geq 0$ is a constant independent of x_1 and x_2.

For example, any function $f : \mathbb{R} \to \mathbb{R}$ with a bounded first derivative must be Lipschitz.

Definition 3.2.9. *A mapping $f : \mathbb{R}^m \to \mathbb{R}^n$ is (k, α)-Hölderian if there exists a real constant $C \geq 0$ such that*

$$\left\| \frac{\partial^{|j|} f}{\partial x^j}(x_2) - \frac{\partial^{|j|} f}{\partial x^j}(x_1) \right\|_{\mathbb{R}^n} \leq C \left(\|x_2 - x_1\|_{\mathbb{R}^m} \right)^\alpha$$

for all $x_1, x_2 \in \mathbb{R}^m$ and for all multi-indices $|j| \leq k$.

Now, the definition of $C^{k,\alpha}$ coordinate charts extends the definition in Example 1.5.8 in the obvious way.

Definition 3.2.10. *A $C^{k,\alpha}$ Hölderian coordinate chart is a one-to-one (k,α)-Hölderian mapping from an open set $\Omega_1 \subseteq \mathbb{R}^m$ to an open set $\Omega_2 \subseteq \mathbb{R}^n$. The set of all $C^{k,\alpha}$ Hölderian coordinate charts from Ω_1 to Ω_2 is denoted by $C^{k,\alpha}(\Omega_1, \Omega_2)$.*

Definition 3.2.11. *Let Ω be a domain in $\mathbb{R}^n$. We say that Ω is (k,α)-Hölderian, or of class $C^{k,\alpha}$, if the conditions (2), (3) and (4) of Definition 3.2.7 are satisfied with a coordinate chart $\phi_x \in C^{k,\alpha}(O_x, B_1(0))$ and whose inverse $\phi_x^{-1} \in C^{k,\alpha}(B_1(0), O_x)$ for some $k \geq 0$ and $\alpha \in (0,1]$.*

Definition 3.2.12. *A Lipschitz domain is a $(0,1)$-Hölderian domain.*

Theorem 3.2.4. *Suppose that Ω is a bounded Lipschitz domain. For positive integers k, r such that $k \geq r \geq 0$, we have the inclusions*

$$W^{k,p}(\Omega) \subseteq W^{r,p}(\Omega)$$

for $1 \leq p \leq \infty$. Moreover, for $k \geq 0$ we have

$$W^{k,p_1}(\Omega) \subseteq W^{k,p_2}(\Omega)$$

whenever $1 \leq p_2 \leq p_1 \leq \infty$.

Of course, this embedding can be interpreted as saying that there is an injection $I : W^{k,p}(\Omega) \to W^{r,p}(\Omega)$ and

$$\|f\|_{W^{r,p}(\Omega)} \overset{\triangle}{=} \|I(f)\|_{W^{r,p}(\Omega)} \leq c\|f\|_{W^{k,p}(\Omega)}.$$

Again while there are numerous types of compactness theorems relating embeddings of Sobolev spaces, the following theorem is prototypical.

Theorem 3.2.5. *Suppose that $\Omega \subseteq \mathbb{R}^n$ is a bounded, Lipschitz domain. The embedding*

$$W^{m,p}(\Omega) \hookrightarrow L^q(\Omega)$$

is continuous and compact for $1 \leq p \leq \infty$, $1 \leq q \leq r$, $m < n/p$ where

$$\frac{1}{r} = \frac{1}{p} - \frac{m}{n}.$$

In addition, the embedding

$$W^{m,p}(\Omega) \hookrightarrow C^0(\overline{\Omega})$$

is continuous and compact for $m > n/p$, $1 \leq p \leq \infty$.

The reader is referred to specialized, focussed treatments of Sobolev and other function spaces in [4, 16, 31] for more complete lists of various embedding theorems.

3.2.4   Trace Theorems

The purpose of this section is to describe in what sense we can prescribe boundary
values to functions in Sobolev spaces. The fundamental problem can be described
as follows: by definition, the Sobolev spaces $W^{s,p}(\Omega)$ are subspaces of $L^p(\Omega)$. The
Sobolev space $W^{s,p}(\Omega)$ consists precisely of those distributions in $L^p(\Omega)$ whose
distributional derivatives up to order $|s|$ are in $L^p(\Omega)$. However, the elements of the
space $L^p(\Omega)$ are equal in norm if they agree almost everywhere, that is except on
a set of Lebesgue measure zero. In other words, $L^p(\Omega)$ is comprised of equivalence
classes of functions. The designation $f \in L^p(\Omega)$ is not meant to refer to a single
function, but to a collection of functions. Two functions g and h are in the same
equivalence class if and only if

$$\|g - h\|_{L^p(\Omega)} = 0.$$

Suppose we take two specific functions in $W^{s,p}(\Omega)$ which differ only on the boundary $\partial\Omega$ of the domain Ω. If the Lebesgue measure of the boundary $\partial\Omega$ is zero as a
subset of $\mathbb{R}^n$, then the two functions are viewed as the same element of $W^{s,p}(\Omega)$.

The construction by which we assign zero boundary values to functions in
$H^m(\Omega)$ is relatively simple to describe. Recall that the space of test functions
$\mathcal{D}$ consists precisely of those smooth functions with support in a compact set
contained in Ω. Thus, if $f \in \mathcal{D}$, it must be equal to zero in a neighborhood of the
boundary. We create a Hilbert space that captures the essence of this construction
by defining $H_0^m(\Omega)$ to be the completion of $\mathcal{D}$ in the $\| \cdot \|_{W^{m,2}(\Omega)}$ norm.

Definition 3.2.13. *The completion of $\mathcal{D}$ in the $\| \cdot \|_{W^{m,2}(\Omega)}$ norm is denoted by
$H_0^m(\Omega)$.*

Functions in $H_0^m(\Omega)$ are thereby interpreted to be functions whose boundary
values on $\partial\Omega$ are zero. At best, this construction provides an intuitive feeling about
how we view the boundary values of functions in $H^m(\Omega)$. The set of results known
as the Trace Theorems gives a precise meaning to what we mean by the boundary
values of functions in H^m. To begin, it is important to note that the Sobolev spaces
$H^m(\Omega)$ and $H_0^m(\Omega)$ differ significantly.

Theorem 3.2.1. *For all integers $m \geq 0$, we have*

$$H_0^m(\Omega) \subseteq H^m(\Omega)$$

and $H_0^m(\Omega)$ is not dense in $H^m(\Omega)$.

Proof. The proof that $H_0^m(\Omega) \subseteq H^m(\Omega)$ follows easily from the following theorem.

Theorem 3.2.6. *For $m \in \mathbb{N}$ and $1 \leq p \leq \infty$, the Sobolev space $W^{m,p}(\Omega)$ is the
completion of*

$$\{f \in C^m(\overline{\Omega}) : \|f\|_{W^{m,p}(\Omega)} < \infty\}$$

in the $\| \cdot \|_{W^{m,p}(\Omega)}$ norm.

A clear, concise proof that $H_0^m(\Omega)$ is not dense in $H^m(\Omega)$ can be found in several places, for example [24]. □

The following theorem shows that not only is $H_0^m(\Omega)$ a *proper subset* of $H^m(\Omega)$, it is the kernel of a map γ, the trace operator, that evaluates smooth functions on the boundary of Ω.

Theorem 3.2.7. *Let Ω be a C^m domain with $m \geq 1$. There is a collection of bounded linear operators $\gamma_j \in \mathcal{L}(H^m(\Omega), H^{m-j-\frac{1}{2}}(\partial\Omega))$ where $j = 0, 1, \ldots, m-1$ such that*

(i)
$$\gamma_j w = \left.\frac{\partial^j w}{\partial n_j}\right|_{\partial\Omega}$$

for all smooth w and $j = 0, 1, \ldots, m-1$, where $\frac{\partial^j}{\partial n_j}$ is the normal derivative of order j on the boundary $\partial\Omega$.

(ii)
$$H_0^m(\Omega) = \{w \in H^m(\Omega) : \gamma_j w = 0 \text{ for } j = 0, 1, \ldots, m-1\}.$$

The implication of this theorem is that the trace operators γ_j agree with our intuition when the function w is smooth enough in a classical sense: they merely evaluate the function or its normal derivative on the boundary in this case. In addition, the trace operators "evaluate to zero" if and only if a function is in $H_0^m(\Omega)$. Thus, the trace operator is used to represent the boundary values for partial differential equations. A formal statement like

$$\int_{\partial\Omega} wv ds, \quad w \in H^1(\Omega), \ v \in L^2(\partial\Omega)$$

should be interpreted as

$$\int_{\partial\Omega} \gamma_0(w) v ds.$$

3.2.5 The Poincaré Inequality

By definition, the Sobolev norm includes terms that contain all derivatives up to a specific order. For example, in $\mathbb{R}^2$, a simple example is

$$\begin{aligned}
\|f\|_{H^1(\Omega)}^2 &= \sum_{\|s\| \leq 1} \|D^s f\|_{L^2(\Omega)}^2 \\
&= \left\|\frac{\partial f}{\partial x_1}\right\|_{L^2(\Omega)}^2 + \left\|\frac{\partial f}{\partial x_2}\right\|_{L^2(\Omega)}^2 + \|f\|_{L^2(\Omega)}^2.
\end{aligned} \tag{3.3}$$

In some important applications, we will be able to obtain simpler expressions for the Sobolev norm when we restrict our attention to those functions that have zero boundary values. More precisely, we will be able to omit certain low-order terms from the summation in Equation (3.3). Obviously, this is not always possible. The essential result relies on an inequality having the form

$$\|f\|_{L^2(\Omega)} \le c\,\|\nabla f\|_{L^2(\Omega)} \tag{3.4}$$

wherever Ω is bounded and $f \in H_0^1(\Omega)$. This result is made precise in the celebrated Poincaré Inequality.

Theorem 3.2.2 (Poincaré Inequality). *Let $\Omega \subseteq \mathbb{R}^n$ be a bounded domain. Then there is a constant c that depends only on the diameter of Ω such that*

$$\|f\|_{H^1(\Omega)} \le c\,\|\nabla f\|_{L^2(\Omega)} \tag{3.5}$$

for all $f \in H_0^1(\Omega)$.

Proof. By definition, $H_0^1(\Omega)$ is the completion of $C_0^\infty(\Omega)$ in the usual norm on $H^1(\Omega)$. We will establish the inequality (3.5) for a function f in $C_0^\infty(\Omega)$, and the theorem will then follow via a standard density and extension by continuity argument. Consider the following identity

$$\begin{aligned}
\|f\|_{L^2(\Omega)}^2 &= \frac{1}{n} \sum_{i=1}^{n} \int_{\Omega} |f(x)|^2 dx \\
&= \frac{1}{n} \sum_{i=1}^{n} \int_{\Omega} (|f(x)|^2) \cdot 1 \, dx.
\end{aligned} \tag{3.6}$$

Because we have assumed that $f \in C_0^\infty(\Omega)$, we can integrate Equation (3.6) by parts in each coordinate direction to obtain

$$\|f\|_{L^2(\Omega)}^2 = \frac{1}{n} \sum_{i=1}^{n} \left\{ \underbrace{|f|^2 x_i \Big|_{\partial \Omega}}_{\to 0} - \int_{\Omega} 2f \frac{\partial f}{\partial x_i} x_i \, dx \right\}.$$

The first term on the right is zero by virtue of the assumption that $f \in C_0^\infty(\Omega)$. By taking the absolute value of each side we obtain the inequality

$$\|f\|_{L^2(\Omega)}^2 \le \frac{2}{n} \sum_{i=1}^{n} \int_{\Omega} \left| \frac{\partial f}{\partial x_i} \right| |f| \, |x_i| \, dx. \tag{3.7}$$

By assumption, the domain Ω is bounded. There is some constant a such that

$$|x_i| < a.$$

As a consequence, the inequality (3.7) becomes

$$\|f\|_{L^2(\Omega)}^2 \le \frac{2a}{n} \sum_{i=1}^{n} \int_{\Omega} \left| \frac{\partial f}{\partial x_i} \right| |f| \, dx.$$

By the Cauchy-Schwarz inequality, we can write

$$\|f\|^2_{L^2(\Omega)} \leq \frac{2a}{n} \sum_{i=1}^{n} \left(\left\| \frac{\partial f}{\partial x_i} \right\|_{L^2(\Omega)} \|f\|_{L^2(\Omega)} \right)$$

or

$$\|f\|_{L^2(\Omega)} \leq \frac{2a}{n} \sum_{i=1}^{n} \left\| \frac{\partial f}{\partial x_i} \right\|_{L^2(\Omega)}. \tag{3.8}$$

We have nearly obtained our desired intermediate result in Equation (3.4). We can rewrite Equation (3.8) as

$$\|f\|_{L^2(\Omega)} \leq \frac{2a}{n} \sum_{i=1}^{n} \left(\int_{\Omega} \left| \frac{\partial f}{\partial x_i} \right|^2 dx \cdot 1 \right)^{1/2}$$

$$= \frac{2a}{n} \sum_{i=1}^{n} \left\{ \left(\int_{\Omega} \left| \frac{\partial f}{\partial x_i} \right|^2 dx \right)^{1/2} \cdot 1 \right\}.$$

If we apply the Cauchy-Schwarz inequality in $\mathbb{R}^n$, we obtain

$$\|f\|_{L^2(\Omega)} \leq \frac{2a}{n} \left(\sum_{i=1}^{n} \left(\int_{\Omega} \left| \frac{\partial f}{\partial x_i} \right|^2 dx \right)^{(1/2) \cdot 2} \right)^{1/2} \cdot \left(\sum_{i=1}^{n} 1^2 \right)^{1/2}$$

$$= \frac{2a}{n} \|\nabla f\|_{L^2(\Omega)} \cdot \sqrt{n}.$$

The result is precisely the form of the inequality in Equation (3.4):

$$\|f\|_{L^2(\Omega)} \leq \frac{2a}{\sqrt{n}} \|\nabla f\|_{L^2(\Omega)}. \tag{3.9}$$

It remains to establish the inequality (3.5). By definition, we have

$$\|f\|^2_{H^1(\Omega)} = \|f\|^2_{L^2(\Omega)} + \|\nabla f\|^2_{L^2(\Omega)}.$$

From Inequality (3.9) we have

$$\|f\|^2_{H^1(\Omega)} \leq \left(1 + \frac{4a^2}{n} \right) \|\nabla f\|^2_{L^2(\Omega)}$$

$$\|f\|_{H^1(\Omega)} \leq \sqrt{1 + \frac{4a^2}{n}} \; \|\nabla f\|_{L^2(\Omega)}$$

$$\|f\|_{H^1(\Omega)} \leq c \|\nabla f\|_{L^2(\Omega)}$$

which is the desired result. Recall that we have established this theorem under the assumption that $f \in C_0^\infty(\Omega)$. The result can be extended for all $f \in H_0^1(\Omega)$ by a standard extension by continuity argument. $\square$

3.3   Banach Space Valued Functions

In the second book of this series, we will review linear first-order and second-order evolution equations. These equations play a central role in applications in engineering and physics. However, they provide not only representations of a host of practical physical problems, but also the foundation for excursions to classes of nonlinear evolution equations.

Simply put, a rigorous treatment of both of these classes of evolution equations absolutely requires that we review fundamentals of Banach space valued functions and their integration in abstract spaces. Fortunately, the construction of integrals of Banach space valued functions is easier to understand when we keep in mind that the construction proceeds, at least strategically, in a manner similar to that employed for real-valued functions.

3.3.1   Bochner Integrals

Let $(X, \mathcal{S}, \mu)$ be a measure space and let Y be a Banach space. A Banach space valued function f

$$f : X \to Y$$

is said to be *finitely-valued* if

- $\mathrm{card}(f(X)) \leq N < \infty$. That is, there exists a finite set $\{y_i\}_{i=1}^N$ such that $f(x) \in \{y_i\}_{i=1}^N$ for all $x \in X$.
- The inverse image of each value in Y is a measurable subset in X. That is

$$B_i = \{x \in X : f(x) = y_i\} \tag{3.10}$$

  is measurable for each $i = 1, \ldots, N$, and
- The class of sets $\{B_i\}_{i=1}^N$ covers X, except possibly for a set of measure zero.

As will be demonstrated shortly, finitely-valued functions in Banach spaces play a role that is analogous to simple functions in the theory of integration of real-valued functions. The Banach space valued function $f : X \to Y$ is strongly Y-measurable if and only if it is the strong limit of finitely-valued functions μ a.e. in X. It is no accident that the definition of the integral of finitely-valued functions resembles the definition of the integral of simple, real-valued functions given in Definition 1.6.9.

Definition 3.3.1. *Let $(X, \mathcal{S}, \mu)$ be a measure space and let $f : X \to Y$ be a finitely-valued function where Y is a Banach space. The integral of f is defined to be*

$$\int_X f(x)\mu(dx) = \sum_{i=1}^N y_i \mu(B_i)$$

where y_i and B_i are defined in Equation (3.10).

Now, we can define the integral of strongly Y-measurable functions. The reader who is unfamiliar with these constructions should note that the integral of Banach space valued functions is defined in terms of finitely-valued Banach space valued functions. This construction bears close resemblance to the construction of integrals of real-valued integrals in terms of simple functions in Definition 1.6.12.

Definition 3.3.2. *Let $(X, \mathcal{S}, \mu)$ be a measure space, and let $f : X \to Y$ where Y is a Banach space. The function f is Bochner integrable if and only if*

1. *there is a sequence of finitely-valued functions f_n such that*

$$f_n \to f \quad \text{a.e. in } X,$$

   and
2. *the limit*

$$\lim_{n \to \infty} \int_X \| f(x) - f_n(x) \|_Y \, \mu(dx) \to 0 \tag{3.11}$$

   as $n \to \infty$.

At this point, we would emphasize that the existence of the Bochner integral relies on the existence of a convergent sequence of finitely-valued functions, and on the existence and convergence of a sequence of integrals of real-valued functions. If the limit in Equation (3.11) exists, we define the Bochner integral to be given by

$$\int_X f(x)\mu(dx) = \lim_{n \to \infty} \int_X f_n(x)\mu(dx) \in Y. \tag{3.12}$$

At least two questions arise in the definition of the Bochner integral in Equation (3.12) above. First, does the condition (3.11) guarantee that the limit in Y expressed in Equation (3.12) exists? In addition, does the limit in (3.12) depend on the sequence of finitely-valued functions $\{f_n\}_{n=1}^{\infty}$? The fact that condition (3.12) does indeed follow from condition (3.11) is evident from the following proposition:

Proposition 3.3.1. *Suppose that $(X, \mathcal{S}, \mu)$ is a measure space, Y is a Banach space, $f : X \to Y$, and $\{f_n\}_{n=1}^{\infty}$ is a sequence of finitely-valued functions such that*

$$f_n \to f \quad \text{a.e. in } X,$$

and

$$\lim_{n \to \infty} \int_X \| f(x) - f_n(x) \|_Y \, \mu(dx) \to 0.$$

Then there is a unique element $F \in Y$ such that

$$F = \lim_{n \to \infty} \int_X f_n(x)\mu(dx).$$

Proof. For each m, $n \in \mathbb{N}$, we can write

$$\left\| \int_X f_n(x)\mu(dx) - \int_X f_m(x)\mu(dx) \right\|_Y \leq \int_X \|f_n(x) - f_m(x)\|_Y \, \mu(dx)$$

because the integral of finitely-valued functions is just a finite sum. By hypothesis, we have

$$\int_X \|f_n(x) - f_m(x)\|_Y \, d\mu(x) \leq \int_X \|f_n(x) - f(x) + f(x) - f_m(x)\|_Y \mu(dx)$$

$$\leq \int_X \|f_n(x) - f(x)\|_Y \, \mu(dx) + \int_X \|f_m(x) - f(x)\|_Y \, \mu(dx)$$

and the right-hand side approaches zero as $m, n \to \infty$. But this means that the sequence $\left\{ \int_X f_n(x)\mu(dx) \right\}_{n=1}^{\infty}$ is a Cauchy sequence in the Banach space Y, and hence has a limit. We define the integral of f to be the limit of the sequence $\{f_n\}_{n \in \mathbb{N}}$,

$$F = \int_X f(x)\mu(dx) \triangleq \lim_{n \to \infty} \int_X f_n(x)\mu(dx). \qquad (3.13)$$

This limit is unique and does not depend on the selection of the sequence $\{f_n\}_{n \in \mathbb{N}}$ of finitely-valued functions. Suppose that $\{g_n\}_{n \in \mathbb{N}}$ and $\{h_n\}_{n \in \mathbb{N}}$ are two sequences of finitely-valued functions such that

$$\|g_n(x) - f(x)\|_Y \to 0 \qquad x \text{ a.e. in } X$$

$$\|h_n(x) - f(x)\|_Y \to 0 \qquad x \text{ a.e. in } X$$

and furthermore

$$\int_X \|f(x) - g_n(x)\|_Y \mu(dx) \to 0$$

$$\int_X \|f(x) - h_n(x)\|_Y \mu(dx) \to 0.$$

From the argument used in deriving Equation (3.13), we know that

$$F_g \triangleq \lim_{n \to \infty} \int g_n(x)\mu(dx)$$

$$F_h \triangleq \lim_{n \to \infty} \int h_n(x)\mu(dx).$$

We construct a sequence $\{f_n(x)\}_{n \in \mathbb{N}}$ by alternating elements of the sequences $\{g_n\}_{n \in \mathbb{N}}$ and $\{h_n\}_{n \in \mathbb{N}}$

$$f_n(x) \triangleq \begin{cases} g_n(x), & \text{if } n = 2k \\ h_n(x), & \text{if } n = 2k - 1 \end{cases} \qquad \text{for } k \in \mathbb{N}.$$

By construction, $\{f_n\}_{n\in\mathbb{N}}$ is a sequence of finitely-valued functions. We have

$$\int_X \|f(x) - f_n(x)\|_Y \mu(dx)$$

$$\leq \max\left\{\int_X \|f(x) - g_n(x)\|_Y \mu(dx), \int_X \|f(x) - h_n(x)\|_Y \mu(dx)\right\}.$$

By the same argument as used in the derivation of Equation (3.13), we can conclude that there is a limit

$$F = \lim_{n\to\infty} \int_X f_n(x)\mu(dx).$$

However, $\left\{\int_X g_n(x)\mu(dx)\right\}_{n\in\mathbb{N}}$ and $\left\{\int_X h_n(x)\mu(dx)\right\}_{n\in\mathbb{N}}$ are just subsequences of the convergent sequence $\left\{\int_X f_n(x)\mu(dx)\right\}_{n\in\mathbb{N}}$. They must converge to the same limit so that

$$F_g = F = F_h$$

and the value of the Bochner integral does not depend on the representation. $\quad\square$

As a final result in this section, the following theorem shows that the Bochner integrable functions are precisely those whose norm is integrable as a real-valued function with respect to the measure μ.

Theorem 3.3.1. *Let $(X, \mathcal{S}, \mu)$ be a measure space and let $f : X \to Y$ where Y is a Banach space. The strongly measurable Banach space valued function f is Bochner integrable if and only if the real-valued function $\|f(x)\|_Y$ is μ-integrable.*

Proof. Throughout the proof, let f be a strongly measurable Banach space valued function, with $\{f_n\}_{n=1}^{\infty}$ a sequence of finitely-valued functions $f_n : X \to Y$ such that

$$f_n \to f \quad \text{a.e. in } X.$$

Suppose that f is Bochner integrable. In this case we know that there is a sequence $\{f_n\}_{n\in\mathbb{N}}$ such that

$$f_n(x) \to f(x) \quad \text{a.e. in } X$$

and

$$\lim_{n\to\infty} \int_X \|f(x) - f_n(x)\|_Y \mu(dx) \to 0.$$

By the triangle inequality, we can write

$$\big|\|f(x)\|_Y - \|f_n(x)\|_Y\big| \leq \|f(x) - f_n(x)\|_Y$$

and conclude that

$$\|f_n(x)\|_Y \to \|f(x)\|_Y \qquad x \text{ a.e. in } X.$$

Additionally, we know that there exists $\gamma \in \mathbb{R}$ such that

$$\gamma = \lim_{n \to \infty} \int_X \|f_n(x)\|_Y \mu(dx)$$

since

$$\left| \int_X \|f_n(x)\|_Y \mu(dx) - \int_X \|f_m(x)\|_Y \mu(dx) \right| \leq \int_X \left| \|f_n(x)\|_Y - f_m(x)\|_Y \right| \mu(dx)$$

$$\leq \int_X \|f_n(x) - f_m(x)\|_Y \mu(dx)$$

$$\leq \int_X \|f_n(x) - f(x)\|_Y \mu(dx) + \int_X \|f_m(x) - f(x)\|_Y \mu(dx).$$

But the right-hand side of the above inequality approaches zero as n approaches infinity by the Bochner integrability of f. Thus the sequence

$$\left\{ \int_X \|f_n(x)\|_Y \mu(dx) \right\}_{n \in \mathbb{N}}$$

is a Cauchy sequence of real numbers, and has a limit $\gamma \in \mathbb{R}$. By Fatou's lemma, $\|f(x)\|_Y$ is Lebesgue integrable satisfying

$$\int_X \|f(x)\|_Y \mu(dx) \leq \liminf_{n \to \infty} \int_X \|f_n(x)\|_Y \mu(dx)$$

$$= \lim_{n \to \infty} \int_X \|f_n(x)\|_Y \mu(dx)$$

$$= \gamma < \infty.$$

Since γ is finite, $\|f(x)\|_Y$ is integrable. On the other hand, suppose that $\|f(x)\|_Y$ is μ-integrable. By hypothesis f is strongly measurable and, consequently, there exists a sequence of finitely-valued functions $\{f_n\}_{n \in \mathbb{N}}$ that converge a.e. in X to f. Define the finitely-valued functions $\{g_n\}_{n=1}^{\infty}$ from the $\{f_n\}_{n=1}^{\infty}$ according to

$$g_n(x) = \begin{cases} f_n(x) & \text{if} \quad \|f_n(x)\|_Y \leq c\|f(x)\|_Y \\ 0 & \text{if} \quad \|f_n(x)\|_Y > c\|f(x)\|_Y \end{cases}$$

for some constant $c > 1$. Clearly, we have that

$$g_n \to f \quad \text{a.e. in } X$$

and we have that the family of functions $\{g_n\}_{n=1}^{\infty}$ is bounded uniformly,

$$\|g_n(x)\|_Y \leq c\|f(x)\|_Y \qquad \text{for all } x \in X.$$

By the triangle inequality, we can write

$$\|f(x) - g_n(x)\|_Y \leq \|f(x)\|_Y + \|g_n(x)\|_Y \leq (1 + c)\|f(x)\|_Y.$$

By Lebesgue's dominated convergence theorem, we have that

$$\int_S \|f(x) - g_n(x)\|_Y \, \mu(dx) \to 0$$

and f is Bochner integrable. $\qquad\square$

3.3.2 The Space $L^p((0,T),X)$

Equipped with the Bochner integral of Banach space valued functions discussed in the last section, we can construct the analogs of Lebesgue spaces of real-valued functions. For example, we define the space of Lebesgue integrable real-valued functions as the completion of the step functions in the L^1 norm. That is, the space of real-valued functions $L^1(0,T)$ can be defined as

$$L^1(0,T) = \left\{ f : \mathbb{R} \to \mathbb{R} \text{ such that } \int_0^T |f| dx < \infty \right\}$$

where the integral is interpreted in the Lebesgue sense. Alternatively, the space $L^1(\mathbb{R})$ can be interpreted as the completion in the $L^1(\mathbb{R})$-norm of the set of step functions $\chi(\mathbb{R})$:

$$L^1(\mathbb{R}) = \text{closure of } \{g \in \chi(\mathbb{R})\} \text{ in } \|\cdot\|_{L^1(\mathbb{R})}.$$

Now, suppose that X is a Banach space. We define

$$L^1((0,T),X) = \left\{ f : (0,T) \to X \text{ such that } \int_0^T \|f(x)\|_X \, dx < \infty \right\}.$$

The space $L^1((0,T),X)$ is a Banach space when equipped with the norm

$$\|f\|_{L^1((0,T),X)} = \int_0^T \|f(x)\|_X \, dx.$$

Again, if $\chi((0,T),X)$ is the space of finitely-valued functions from $(0,T)$ into X, we can also define

$$L^1((0,T),X) = \text{closure of } \{g \in \chi((0,T),X)\} \text{ in } \|\cdot\|_{L^1((0,T),X)}.$$

The generalization of the L^p spaces of real-valued functions to Banach space valued functions is immediate. If we consider all Banach space valued functions $f : (0,T) \to X$ such that

$$\int_0^T \|f(x)\|_X^p \, dx < \infty$$

we obtain $L^p((0,T),X)$. The space $L^p((0,T),X)$ is a Banach space when equipped with the norm

$$\|f\|_{L^p((0,T),X)} = \left(\int_0^T \|f(x)\|_X^p \, dx \right)^{1/p}.$$

We have the following important theorem that summarizes two useful characteristics of Banach space valued functions.

Theorem 3.3.2. *Let X, Y be Banach spaces with X continuously embedded in Y.*

- *The embedding*

$$L^p((0,T),X) \subseteq L^q((0,T),Y), \quad 1 \le q \le p \le \infty$$

  is continuous.

- *Let X be a separable and reflexive Banach space, $1 < p < \infty$ and $\frac{1}{p} + \frac{1}{q} = 1$. Then*

$$\left(L^p((0,T),X) \right)^* = L^q((0,T),X^*).$$

Proof. The proof of the first part of this theorem is straightforward when we recall the definition of a continuous embedding. The Banach space X is continuously embedded in Y if and only if

$$\|I(f)\|_Y \le c \|f\|_X$$

for some constant c, where $I : X \to Y$ is the embedding map

$$I : X \quad \to \quad Y$$
$$I : f \in X \quad \mapsto \quad f \in Y.$$

Suppose that $f \in L^p((0,T),X)$ where $1 \le p \le \infty$. We can write

$$\|f\|_{L^p((0,T),X)}^p = \int_0^T \|f(x)\|_X^p \, dx.$$

But we know that

$$\|f(x)\|_Y^q \le \|f(x)\|_Y^p \le c^p \|f(x)\|_X^p.$$

Integrating from 0 to T, we see that

$$\|f\|_{L^p((0,T),Y)}^p \le c^p \|f\|_{L^p((0,T),X)}^p$$

and that

$$\|f\|_{L^q((0,T),Y)}^q = \int_0^T \|f(x)\|_Y^q \, dx \le \int_0^T \|f(x)\|_Y^p \, dx = \|f\|_{L^p((0,T),Y)}^p.$$

The last inequality can be re-written

$$\|f\|_{L^q((0,T),Y)}^q \le c^p \|f\|_{L^p((0,T),X)}^p$$

and the theorem follows. The reader is referred to [37, p. 411] for a proof of the second assertion. $\qquad \square$

3.3.3 The Space $W^{p,q}((0,T), X)$

One of the first topics in many applied functional analysis manuscripts introduces the idea of the generalized derivative of a real-valued function. This notion of differentiation is critical in the definition of the Sobolev scale of weakly differentiable functions. Essentially, a real-valued function f is said to have a weak derivative, or generalized derivative, if there is a function g such that

$$\int_\Omega f(x)\phi'(x)dx = -\int_\Omega g(x)\phi(x)dx \quad \forall \phi \in C_0^\infty(\Omega). \qquad (3.14)$$

That is, we define the weak derivative to be given by the integration by parts formula in Equation (3.14). If this condition holds for all smooth functions with compact support contained in Ω, then by definition $g = f'$ in the generalized, or distributional sense. Obviously, if the classical derivative of f exists, the generalized derivative is equal to the classical derivative.

Now, just as in the last section, we can lift this definition to the framework of Banach space valued functions.

Definition 3.3.3. *Let X be a Banach space, and let $f, g \in L^1((0,T), X)$. We say that g is the nth-order generalized derivative of f if and only if*

$$\int_0^T g(x)\phi(x)dx = (-1)^n \int_0^T f(x)\phi^{(n)}(x)dx$$

for all $\phi \in C_0^\infty(0,T)$, where the integrals are understood in the Bochner sense.

In our treatments of linear first-order and second-order partial differential equations, the generalized derivatives arise within the framework of Gelfand triples of spaces. In this setting, there is a characterization of the generalized derivative that seems particularly well suited for many applications that appear in this book.

Theorem 3.3.3. *Let $V \hookrightarrow H \hookrightarrow V^*$ define a Gelfand triple, and suppose that $f \in L^p((0,T), V)$ and $1 \leq p, q \leq \infty$. There is an nth-order generalized derivative of f,*

$$f^{(n)} \in L^q((0,T), V^*)$$

if and only if there is a function $g \in L^q((0,T), V^*)$ such that

$$\int_0^T (f(x),\xi)_H \phi^{(n)}(x)dx = (-1)^n \int_0^T \langle g(x),\xi\rangle_{V^* \times V}\phi(x)dx \qquad (3.15)$$

for all $\xi \in V$ and for all $\phi \in C_0^\infty(0,T)$. In this case, we define $f^{(n)} = g$ and write

$$\frac{d^{(n)}}{dx^n}(f(x),\xi)_H = \langle g(x),\xi\rangle_{V^* \times V} = \left\langle f^{(n)},\xi\right\rangle_{V^* \times V} \quad \text{a.e. in } x \in [0,T]$$

for all $\xi \in V$. This derivative is understood to be the generalized derivative of the real-valued function $(f(x),\xi)_H$.

Because the duality pairing on $V^* \times V$ can be viewed as the extension by continuity of the H-inner product $(\cdot, \cdot)_H$ from $H \times V$ to $V^* \times V$, the left-hand side of Equation (3.15) can be written as

$$\int_0^T \langle f(x), \xi \rangle_{V^* \times V} \phi^{(n)}(x) dx = \int_0^T (f(x), \xi)_H \phi^{(n)}(x) dx. \qquad (3.16)$$

Now, the following proposition is required to show that the "integral of the duality pairing" is equal to the "duality pairing of the integrals" that comprise the Equation (3.16).

Proposition 3.3.2. *Let V be a Banach space. If $f \in L^p((0, T), V)$ then we have*

$$\int_0^T \langle g, f(x) \rangle_{V^* \times V} dx = \left\langle g, \int_0^T f(x) dx \right\rangle_{V^* \times V} \qquad \forall g \in V^*. \qquad (3.17)$$

If, on the other hand, $g \in L^p((0, T), V^*)$ then we have

$$\int_0^T \langle g(x), f \rangle_{V^* \times V} dx = \left\langle \int_0^T g(x) dx, f \right\rangle_{V^* \times V} \qquad \forall f \in V. \qquad (3.18)$$

Proof. We will only consider the proof of the assertion in Equation (3.17) above, as the proof of the second statement in Equation (3.18) is nearly identical. If $f \in L^p((0, T), V)$, we know that there is a sequence of finitely-valued functions $\{f_n\}_{n=1}^{\infty}$ such that

$$f_n(x) \quad \rightarrow \quad f(x) \quad \text{in } V, \qquad \text{a.e. in } x \in [0, T]$$
$$\int_0^T f_n(x) dx \quad \rightarrow \quad \int_0^T f(x) dx \quad \in V.$$

Since each f_n is finitely-valued, we have

$$\int_0^T \langle g, f_n(x) \rangle_{V^* \times V} dx = \left\langle g, \int_0^T f_n(x) dx \right\rangle_{V^* \times V}. \qquad (3.19)$$

Each of the above integrals is, in fact, a finite sum for each $n \in \mathbb{N}$. The right-hand side of Equation (3.19) converges

$$\left\langle g, \int_0^T f_n(x) dx \right\rangle_{V^* \times V} \rightarrow \left\langle g, \int_0^T f(x) dx \right\rangle_{V^* \times V}$$

because of the strong convergence of the sequence

$$\int_0^T f_n(x) dx \rightarrow \int_0^T f(x) dx \quad \text{in } V.$$

For the term on the left, we can write

$$\underbrace{\left| \int_0^T \langle g, f_n(x) \rangle_{V^* \times V} dx - \int_0^T \langle g, f_m(x) \rangle_{V^* \times V} dx \right|}_{I_1}$$

$$\leq \int_0^T \|g\|_V^* \|f_n(x) - f_m(x)\|_V \, dx.$$

But the right-hand side of the above inequality approaches zero since f is Bochner integrable. Thus, the sequence $\left\{ \int_0^T \langle g, f_n(x) \rangle_{V^* \times V} dx \right\}_{n \in \mathbb{N}}$ is a Cauchy sequence of real numbers and therefore convergent. Since we can write

$$\left| \left| \int_0^T \langle g, f(x) \rangle_{V^* \times V} dx - \int_0^T \langle g, f_n(x) \rangle_{V^* \times V} dx \right| \right.$$

$$\left. - \left| \int_0^T \langle g, f(x) \rangle_{V^* \times V} dx - \int_0^T \langle g, f_m(x) \rangle_{V^* \times V} dx \right| \right| \leq I_1$$

it is clear that $\int_0^T \langle g, f_n(x) \rangle_{V^* \times V} dx \to \int_0^T \langle g, f(x) \rangle_{V^* \times V} dx$. The proposition is therefore proved. $\qquad \square$

Now using this proposition, we can combine equations (3.15) and (3.16) to obtain

$$\left\langle \int_0^T f(x) \phi^{(n)}(x) dx, \xi \right\rangle_{V^* \times V} = \left\langle (-1)^n \int_0^T g(x) \phi(x) dx, \xi \right\rangle_{V^* \times V}$$

which implies that

$$\int_0^T f(x) \phi^{(n)}(x) dx = (-1)^n \int_0^T g(x) \phi(x) dx \quad \forall \phi \in C_0^\infty(0, T).$$

This is the desired result.

Having defined the spaces

$$L^p((0, T), V) \qquad 1 < p < \infty$$

and its topological dual

$$L^q((0, T), V^*)$$

where

$$\frac{1}{p} + \frac{1}{q} = 1$$

we define Sobolev spaces that are comprised of Banach space valued functions. These spaces are crucial to the development of evolution equations.

Definition 3.3.4. *Suppose that $V \hookrightarrow H \hookrightarrow V^*$ forms a Gelfand triple and let $1 < p < \infty$ and $\frac{1}{p} + \frac{1}{q} = 1$. We define*

$$W^{1,p}((0,T), V, H) = \left\{ u \in L^p((0,T), V) : \frac{du}{dt} \in L^q((0,T), V^*) \right\}.$$

The space $W^{1,p}((0,T), V, H)$ is a Banach space when equipped with the norm

$$\|f\|_{W^{1,p}((0,T),V,H)} = \|f\|_{L^p((0,T),V)} + \left\| \frac{df}{dt} \right\|_{L^q((0,T),V^*)}.$$

The embedding
$$W^{1,p}((0,T), V, H) \subseteq C([0,T], H)$$

is continuous.

Chapter 4

Differential Calculus
in Normed Vector Spaces

4.1 Differentiability of Functionals

In our first extension of the notion of differentiability of real-valued functions, we introduce the Gateaux differential of a functional mapping a vector space into a normed vector space. The Gateaux differential will play a crucial role throughout the remainder of this text, and the student is advised to study the definition carefully.

4.1.1 Gateaux Differentiability

Definition 4.1.1. *If A is a function mapping a vector space X into the normed vector space Y, the Gateaux differential $D_G A$ of A at $u \in X$ in the direction $v \in X$ is given by*

$$D_G A(u, v) \equiv \lim_{\epsilon \to 0} \frac{A(u + \epsilon v) - A(u)}{\epsilon}$$

when the limit exists.

It should be noted that, strictly speaking, the Gateaux differential does not require a norm on the vector space X for Definition 4.1.1 to make sense. If the Gateaux differential exists, the Gateaux derivative need not exist. The existence of the Gateaux derivative requires the existence of the Gateaux differential. But it also requires that the derivative defines an operator that is bounded and linear in *the direction* the derivative is taken.

Definition 4.1.2. *Suppose A is a function mapping a vector space X into the normed vector space Y. A function $D_G A : X \times X \to Y$ is the Gateaux derivative at $u \in X$ in the direction $v \in X$ if*

(i) *the Gateaux differential*

$$D_G A(u, v) = \lim_{\epsilon \to 0} \frac{A(u + \epsilon v) - A(u)}{\epsilon},$$

exists for all $v \in X$,

(ii) $D_G A(u, \cdot)$ *is linear, and*
(iii) *the mapping* $D_G A(u, \cdot)$ *is bounded.*

It should be emphasized that the mapping A acts from a normed vector space X to some general normed vector space Y

$$A : X \to Y$$

and the Gateaux differential $D_G A(\cdot, \cdot)$ maps

$$D_G A(\cdot, \cdot) : X \times X \to Y$$

For any fixed $u \in X$, the mapping,

$$D_G A(u, \cdot) : X \to Y$$

is linear and bounded. We will frequently employ the notation

$$D_G A(u) \in \mathcal{L}(X, Y).$$

To define $D_G A(u)$, we equate the mappings

$$v \mapsto D_G A(u, v)$$

and

$$v \mapsto D_G A(u) \circ v.$$

By the hypotheses that $D_G A(u, \cdot)$ is linear and bounded on X, we obtain $D_G A(u) \in \mathcal{L}(X, Y)$. This fact follows since

$$\|D_G A(u) \circ v\|_Y = \|D_G A(u, v)\|_Y \leq c\|v\|_X$$

and it is clear that

$$D_G A(u) \circ (\alpha v + \beta w) = \alpha D_G A(u) \circ v + \beta D_G A(u) \circ w.$$

We are particularly interested in the case in which A is a mapping from the vector space X into $\mathbb{R}$. That is, $Y \equiv \mathbb{R}$ and A is a functional. In this case, the condition that $D_G A(u, \cdot)$ is bounded and linear in X implies that

$$\|D_G A(u, v)\|_Y \leq c\|v\|_X$$

which reduces to

$$|D_G A(u, v)| \leq c\|v\|_X.$$

It is not difficult to show that we can define an operator

$$D_G A(u) \equiv D_G A(u, \cdot) \quad \forall u \in X$$

where $D_G A(u) \in X^*$ for each $u \in X$. In other words, we have that

$$D_G A(u) \in \mathcal{L}(X, \mathbb{R})$$

and the Gateaux derivative can be expressed using the duality pairing

$$\langle D_G A(u), v \rangle_{X^* \times X} = D_G A(u, v) \quad \forall u, v \in X.$$

This representation of the Gateaux derivative will be used extensively in applications, discussed in the second book of this series.

4.1.2   Fréchet Differentiability

Careful inspection of the definition of the Gateaux derivative shows that it can
be defined without any topology on the domain of the operator A. A somewhat
stronger notion of differentiability, one that is well defined for operators mapping
a normed vector space into another normed vector space, is the Fréchet derivative.

Definition 4.1.3. *Let X, Y be normed vector spaces, and suppose that*

$$A : X \rightarrow Y.$$

*A function $D_F A : X \times X \rightarrow Y$ is the Fréchet derivative at $u \in X$ in the direction
$v \in X$ if*

(i) *the limit*

$$\lim_{\|v\| \to 0} \frac{\|f(u+v) - f(u) - D_F A(u,v)\|_Y}{\|v\|_X} = 0$$

exists,

(ii) *the mapping $D_F A(u, \cdot)$ is linear, and*

(iii) *the mapping $D_F A(u, \cdot)$ is bounded.*

As in our discussion of the Gateaux derivative in the previous section, the
same arguments allow us to identify the Fréchet derivative with a bounded linear
operator. We can write

$$\begin{aligned}
D_F A(u) \circ v &\equiv D_F A(u, v) \qquad \forall\, v \in X \\
D_F A(u) &\in \mathcal{L}(X, Y)
\end{aligned}$$

for a fixed $u \in X$. Again, in our most frequently studied applications (in the next
book of this series), we will focus on mappings that are functionals, and the Fréchet
derivative can be expressed in terms of the duality pairing

$$D_F A(u, v) = \langle D_F A(u), v \rangle_{X^* \times X} \qquad \forall\, u, v \in X.$$

As an immediate consequence of the definition for the Fréchet derivative, we have
a type of "Taylor formula" for Fréchet differentiable functions.

Proposition 4.1.1. *Let X, Y be normed vector spaces, and let $f : X \rightarrow Y$ be Fréchet
differentiable at $u \in X$ in the direction $v \in X$. We have*

$$f(u+v) = f(u) + D_F f(u, v) + R(u, v)$$

where the remainder R satisfies

$$\lim_{\|v\|_X \to 0} \frac{\|R(u, v)\|_Y}{\|v\|_X} \to 0.$$

The proof of this proposition is a direct consequence of part (i) in Definition 4.1.3.

One of the most important properties of the Fréchet derivative, in contrast to the Gateaux derivative, is the existence of a chain rule.

Theorem 4.1.1. *Suppose that X, Y and Z are normed vector spaces and that f, g are Fréchet differentiable on open sets O_x and O_y, contained in X and Y, respectively,*

$$g : O_x \subseteq X \to Y$$

$$f : O_y \subseteq Y \to Z.$$

Then the Fréchet derivative of the composition $f \circ g$ is given by the chain rule

$$D_F(f \circ g)(u, v) = D_F f(g(u), D_F g(u, v)).$$

Proof. To prove this theorem, we first note that by definition

$$\begin{aligned} D_F f(\cdot, \cdot) : Y \times Y &\to Z \\ D_F g(\cdot, \cdot) : X \times X &\to Y. \end{aligned}$$

Thus, the expression

$$D_F f(g(\underbrace{u}_{\in X}), \underbrace{D_F g\,(u, v)}_{\in X \times X})$$

makes sense, if the Gateaux differentials $D_F f$ and $D_F g$ exist. Now, $f \circ g : X \to Z$ and

$$D_F(f \circ g) : X \times X \to Z.$$

Consider the difference

$$(f \circ g)(u + v) - (f \circ g)(u).$$

We can directly calculate that

$$\begin{aligned} (f \circ g)(u + v) - (f \circ g)(u) &= f(g(u + v)) - f(g(u)) \\ &= f(\underbrace{g(u + v) - g(u)}_{D_F g(u,v) + R_1(u,v)} + g(u)) - f(g(u)) \end{aligned}$$

since g is Fréchet differentiable. Furthermore, we can write

$$\begin{aligned} (f \circ g)(u + v) - (f \circ g)(u) &= f(g(u) + D_F g(u, v) + R_1(u, v)) - f(g(u)) \\ &= D_F f(g(u), D_F g(u, v) + R_1(u, v)) \\ &\quad + R_2(g(u), D_F g(u, v) + R_1(u, v)) \\ &= D_F f(g(u), D_F g(u, v)) + D_F f(g(u), R_1(u, v)) \\ &\quad + R_2(g(u), D_F g(u, v) + R_1(u, v)) \qquad (4.1) \end{aligned}$$

by the Fréchet differentiability of f. By the definition of the Fréchet derivative, we have

$$\frac{\|R_1(u,v)\|_Y}{\|v\|_X} \to 0 \quad \text{as} \quad \|v\|_X \to 0$$

and

$$\frac{\|R_2(g(u), D_F g(u,v) + R_1(u,v))\|_Z}{\|D_F g(u,v) + R_1(u,v)\|_Y} \to 0$$

as

$$\|D_F g(u,v) + R_1(u,v)\|_Y \to 0.$$

We can write

$$\frac{\|D_F f(g(u), R_1(u,v))\|_Z}{\|v\|_X} \leq c \frac{\|R_1(u,v)\|_Y}{\|v\|_X} \to 0 \tag{4.2}$$

as $\|v\|_X \to 0$, by Definition 4.1.3. By the boundedness of the Fréchet derivative in its second argument, we have

$$\|D_F g(u,v) + R_1(u,v)\|_Y \leq c\|v\|_X + \|R_1(u,v)\|_Y \to 0$$

as $\|v\|_X \to 0$. Finally, we can write

$$\frac{\|R_2(g(u), D_F g(u,v) + R_1(u,v))\|_Z}{\|v\|_X} = \frac{\|R_2(g(u), D_F g(u,v) + R_1(u,v))\|_Z}{\|D_F g(u,v) + R_1(u,v)\|_Y} \cdot \frac{\|D_F g(u,v) + R_1(u,v)\|_Y}{\|v\|_X}$$

$$\leq \underbrace{\frac{\|R_2(g(u), D_F g(u,v) + R_1(u,v))\|_Z}{\|D_F g(u,v) + R_1(u,v)\|_Y}}_{\to 0} \left(c + \underbrace{\frac{\|R_1(u,v)\|_Y}{\|v\|_X}}_{\to 0} \right)$$

$$\tag{4.3}$$

as $\|v\|_X \to 0$. The result now follows when we take the norm of Equation (4.1), divide by $\|v\|_X$, and employ Equations (4.2) and (4.3). $\qquad\square$

Every Fréchet differentiable functional is Gateaux differentiable, but the converse in not true. There are some simple conditions however, which taken together with Gateaux differentiability, are sufficient to guarantee that a functional is Fréchet differentiable. As a preliminary result in establishing this fact, we show that a mean value theorem can be derived for Gateaux (or Fréchet) differentiable functionals. This result is important in its own right.

Theorem 4.1.2. *Let X and Y be linear topological spaces, let U be an open subset of X, and let a mapping*

$$F : U \to Y$$

be Gateaux differentiable at every point of the interval

$$[x, x+h] \subset U.$$

Then,

a) *If the $z \mapsto D_G F(z)h$ is a continuous mapping of the interval $[x, x+h]$ into Y, then*

$$F(x+h) - F(x) = \int_0^1 D_G F(x+th) \circ h dt.$$

b) *If moreover X and Y are Banach spaces, then*

$$\|F(x+h) - F(x)\| \leq \sup_{0 \leq t \leq 1} \|D_G F(x+th)\| \, \|h\|$$

and for any $\Lambda \in \mathcal{L}(X, Y)$,

$$\|F(x+h) - F(x) - \Lambda h\| \leq \sup_{0 \leq t \leq 1} \|D_G F(x+th) - \Lambda\| \, \|h\|.$$

In particular

$$\|F(x+h) - F(x) - D_G F(z)h\| \leq \sup_{0 \leq t \leq 1} \|D_G F(x+th) - D_G F(z)\| \, \|h\|$$

for any point $z \in [x, x+h]$.

Proof. Define $\phi(t) \triangleq F(x+th)$. The definition of the Gateaux derivative yields

$$
\begin{aligned}
\phi'(t) &= \lim_{\epsilon \to 0} \frac{\phi(t+\epsilon) - \phi(t)}{\epsilon} \\
&= \lim_{\epsilon \to 0} \frac{F(x + (t+\epsilon)h) - F(x+th)}{\epsilon} \\
&= D_G F(x+th) \circ h.
\end{aligned}
$$

By the mean value theorem for the real variable function ϕ

$$
\begin{aligned}
\phi(1) - \phi(0) &= \int_0^1 \phi'(t) dt \\
F(x+h) - F(x) &= \int_0^1 D_G F(x+th) \circ h dt.
\end{aligned}
$$

Suppose that $\Lambda \in \mathcal{L}(X, Y)$. By direct computation, we have

$$
\begin{aligned}
\|F(x+h) - F(x) - \Lambda h\| &= \left\| \int_0^1 D_G F(x+th) \circ h dt - \Lambda h \right\| \\
&= \left\| \int_0^1 (D_G F(x+th) - \Lambda) \circ h dt \right\| \\
&\leq \int_0^1 \left\| (D_G F(x+th) - \Lambda) \circ h \right\| dt \\
&\leq \int_0^1 \left\| (D_G F(x+th) - \Lambda) \right\| \|h\| dt \\
&\leq \sup_{t \in [0,1]} \|D_G F(x+th) - \Lambda\| \, \|h\|. \qquad \square
\end{aligned}
$$

With the mean value theorem, we can derive useful conditions that guarantee that a given Gateaux differentiable functional is in fact Fréchet differentiable.

Corollary 4.1.1. *Let X be a Banach space and let F be a continuous mapping of a neighborhood $U \subset X$ of a point $x_0 \in X$ into a Banach space Y. Assume that the mapping is Gateaux differentiable at every point of U, and that the mapping $x \mapsto D_G F(x)$ from U into $\mathcal{L}(X, Y)$ (considered with the uniform operator topology) is continuous. Then F is Fréchet differentiable on U and for all $x \in U$*

$$D_G F(x) = D_F F(x).$$

Proof. By the mean value theorem

$$\|F(x_0 + h) - F(x_0) - D_G F(x_0) \circ h\| \le \sup_{0 \le t \le 1} \|D_G F(x_0 + th) - D_G F(x_0)\| \, \|h\|.$$

Since the mapping $x \mapsto D_G F(x)$ is continuous, for every $\epsilon > 0$ there exists $\delta > 0$ such that

$$\|x - x_0\|_X < \delta \qquad \Longrightarrow \qquad \|D_G F(x) - D_G F(x_0)\|_Y < \epsilon.$$

Therefore

$$\lim_{\|h\|_X \to 0} \frac{\|F(x_0 + h) - F(x_0) - D_G F(x_0) \circ h\|_Y}{\|h\|_X} = 0$$

which means that F is Fréchet differentiable and

$$D_G F(x) = D_F F(x). \qquad \square$$

4.2 Classical Examples of Differentiable Operators

The following examples, while elementary, serve as a conceptual bridge between classical derivatives as studied in calculus, and those differentials that must be calculated routinely in practical problems of mechanics and control applications. In this first example, we show that affine functions on an abstract space are Fréchet differentiable. The derivative agrees with our intuition that a derivative of an affine function "picks off" the coefficient of the independent variable.

In examples, we omit indices G and F at differential D, since it will be clear from context whether the Gateaux or Fréchet differential is considered.

Example 4.2.1. *Suppose X, Y are normed vector spaces and*

$$T : X \to Y$$
$$T(x) \triangleq Ax + y$$

where A is a bounded linear operator from X into Y, and $y \in Y$ is fixed. Then T is Fréchet differentiable, and hence Gateaux differentiable, at all $x \in X$. By the linearity of A, we have

$$T(x + h) - T(x) = Ah.$$

For any $x \in X$, define

$$DT(x) \circ h \triangleq Ah$$

then clearly

$$\lim_{h \to \infty} \frac{\left\| T(x+h) - T(x) - DT(x) \circ h \right\|_Y}{\|h\|_X} = 0.$$

By inspection, the Fréchet derivative at any fixed $x \in X$ is a bounded linear operator acting on $h \in X$. We can compute

$$\left\| DT(x) \circ h \right\|_Y = \|Ah\|_Y \le C\|h\|_X$$

by the boundedness of the operator A.

In this example we show that a classically differentiable, real-valued function defined on the real line is both Gateaux and Fréchet differentiable, provided some rather weak assumptions hold. The example again serves to emphasize that differentiation of functionals extends our conventional notion of differentiation.

Example 4.2.2. *Let $f : O \subset \mathbb{R} \to \mathbb{R}$ be a classically differentiable real-valued function defined on the open set O. Let us calculate the Gateaux differential of f at $x_0 \in O$ in the direction $y \in \mathbb{R}$. By definition,*

$$Df(x_0, y) = \lim_{\epsilon \to 0} \frac{f(x_0 + \epsilon y) - f(x_0)}{\epsilon}.$$

For $y \neq 0$, we can always write

$$
\begin{aligned}
\lim_{\epsilon \to 0} \frac{f(x_0+\epsilon y)-f(x_0)}{\epsilon} &= \lim_{\epsilon \to 0} \frac{f(x_0+\epsilon y)-f(x_0)}{\epsilon y} y \\
&= \lim_{\epsilon' \to 0} \frac{f(x_0+\epsilon')-f(x_0)}{\epsilon'} y \\
&= \frac{df}{dx}(x_0)y.
\end{aligned}
$$

Hence, we have that the Gateaux differential is given by

$$Df(x_0, y) = \frac{df}{dx}(x_0)y.$$

Let us show that f is also Gateaux differentiable. According to the definition, we must also show that $Df(x_0, \cdot)$ is linear and bounded. Linearity in y is clear. Boundedness follows if the classical derivative is bounded over the open set O. This condition holds if $f \in C^1(\Omega)$. We should also note that in this case since $f : O \subset \mathbb{R} \to \mathbb{R}$, we can alternatively write

$$Df(x_0, y) = \langle Df(x_0), y \rangle_{X^* \times X}.$$

In the next example, we show that the results for the derivative of a function acting from $\mathbb{R}$ to $\mathbb{R}$ can be extended readily to multiple dimensions. The analysis requires, of course, the introduction of the classical derivative of functions on $\mathbb{R}^n$.

Example 4.2.3. *Let $f : O \subset \mathbb{R}^n \to \mathbb{R}$ be a real-valued, classically differentiable function defined on the open set O. Let us calculate the Gateaux differential of f at $x_0 \in O$ in the direction $y \in \mathbb{R}^n$. As in the last example, the answer is nearly immediate, given that f satisfies such strong classical differentiability properties. Indeed, by definition we have*

$$Df(x_0, y) = \lim_{\epsilon \to 0} \frac{f(x_0 + \epsilon y) - f(x_0)}{\epsilon}.$$

To calculate the Gateaux differential of $f : O \subseteq \mathbb{R}^n \to \mathbb{R}$, let us recall the definition of the classical derivative of a function $g : \mathbb{R}^n \to \mathbb{R}^m$.

Definition 4.2.1. *Suppose $g : O \subseteq \mathbb{R}^n \to \mathbb{R}^m$ where O is an open set. The function g is classically differentiable at $x_0 \in O$ if*

(i) *The partial derivatives of g, $\frac{\partial g_i}{\partial x_j}$ for $i = 1, \ldots, m$ and $j = 1, \ldots, n$ exist at x_0, and*

(ii) *The Jacobian matrix $J(x_0) = \left[\frac{\partial g_i}{\partial x_j}(x_0) \right] \in \mathbb{R}^{m \times n}$ satisfies*

$$\lim_{x \to x_0} \frac{\| g(x) - g(x_0) - J(x_0)(x - x_0) \|}{\| x - x_0 \|} = 0. \tag{4.4}$$

We say that the Jacobian matrix $J(x_0)$ is the derivative of g at x_0.

To return to our construction of the Gateaux derivative from Equation (4.4), we can write from the definition of the classical derivative

$$\lim_{\epsilon \to 0} \frac{\| f(x_0 + \epsilon y) - f(x_0) \|}{\| \epsilon y \|} = \lim_{\epsilon \to 0} \frac{\| J(x_0)(\epsilon y) \|}{\| \epsilon y \|}$$

$$= \frac{\| J(x_0) y \|}{\| y \|}$$

$$\lim_{\epsilon \to 0} \frac{\| f(x_0 + \epsilon y) - f(x_0) \|}{\epsilon} = \frac{\| J(x_0) y \|}{\| y \|} \| y \|$$

and consequently

$$Df(x_0) \circ y = J(x_0) y.$$

Again, it is not difficult to show that f is in fact Gateaux differentiable and Fréchet differentiable. Linearity in $y \in \mathbb{R}^n$ is clear. Boundedness follows since

$$\| Df(x_0, y) \| = \| Df(x_0) \circ y \|_{m,1} \leq \| J(x_0) \|_{m,n} \| y \|_{n,1}$$

where $\| \cdot \|_{m,n}$ is a matrix norm. Hence, we have

$$Df(x_0) = J(x_0).$$

Example 4.2.4. *Let $f : C^\infty(K) \to \mathbb{R}$ be a function defined via the integral operator*

$$f(x) = \int_K x(\xi)d\xi$$

where $x \in C^\infty(K)$ is a smooth function and K is a closed, bounded subset of $\mathbb{R}$. We calculate the Gateaux differential of f at $x_0 \in C^\infty(K)$ in the direction $y \in C^\infty(K)$. Note that in contrast to the previous two examples, x is a function, not an element of the real line. Similarly, while f maps into $\mathbb{R}$, its domain is a collection of smooth functions. We apply the definition directly

$$\begin{aligned}
Df(x_0, y) &= \lim_{\epsilon \to 0} \frac{f(x_0 + \epsilon y) - f(x_0)}{\epsilon} \\
&= \lim_{\epsilon \to 0} \frac{1}{\epsilon}\left\{ \int_K (x_0(\xi) + \epsilon y(\xi))d\xi - \int_K x_0(\xi)d\xi \right\} = \int_K y(\xi)d\xi.
\end{aligned}$$

Obviously, this expression for $Df(x_0, y)$ is linear in y. We have

$$Df(x_0, y + \alpha z) = Df(x_0, y) + \alpha Df(x_0, z).$$

But is this expression bounded in its second argument? Recall that

$$Df(\cdot, \cdot) : C^\infty(K) \times C^\infty(K) \to \mathbb{R}.$$

We require the map

$$Df(x_0, \cdot) : C^\infty(K) \to \mathbb{R}$$

to be a bounded transformation. According to the definition, this means that we must find a constant c such that

$$\|Df(x_0, y)\| = |Df(x_0, y)| \le c\|y\|.$$

By definition

$$C^\infty(K) \overset{\triangle}{=} \bigcap_{k=0}^{\infty} C^k(K) \subseteq C(K).$$

So, $C^\infty(K)$ is a subspace of $C(K)$ and we can endow it with the norm $\|\cdot\|_{C(K)}$. In other words, we will consider that $C^\infty(K)$ is endowed with the norm

$$\|y\| = \sup_{\xi \in K}|y(\xi)|.$$

We can consequently write

$$\begin{aligned}
|Df(x_0, y)| &= \left| \int_K y(\xi)d\xi \right| \le \int_K |y(\xi)|d\xi \\
&\le \int_K \sup_{\xi \in K}|y(\xi)|ds \le \mathrm{meas}(K)\|y\|_{C(K)}
\end{aligned}$$

which is the desired result.

The last example provided an introduction to functional differentiation on abstract spaces, those different from $\mathbb{R}^n$. The next example is similar in nature, but finds more application in practice.

Example 4.2.5. *Consider the mapping* $F : C[t_0, t_1] \to \mathbb{R}$ *where*

$$F(g) \triangleq \int_{t_0}^{t_1} g(s)ds.$$

We claim that F *is Fréchet differentiable and*

$$DF(g) \circ h = \int_{t_0}^{t_1} h(s)ds.$$

In fact, this result is nearly immediate. We can write

$$\|F(g+h) \quad -F(g) - Df(g) \circ h\|$$
$$= \left\| \int_{t_0}^{t_1} (g(s) + h(s))\, ds - \int_{t_0}^{t_1} g(s)ds - \int_{t_0}^{t_1} h(s)ds \right\|$$
$$= 0.$$

Moreover

$$\|DF(g) \circ h\| = \left\| \int_{t_0}^{t_1} h(s)ds \right\|$$
$$\leq \int_{t_0}^{t_1} \|h(s)\|ds$$
$$\leq \int_{t_0}^{t_1} \sup_{\tau \in [t_0, t_1]} |h(\tau)|ds$$
$$= (t_1 - t_0)\|h\|_{C[t_0,t_1]}$$

so that $DF(g)$ *is a bounded and linear functional on* $C[t_0, t_1]$.

Quite complicated examples can be built up from much simpler building blocks. The following example is an instructive exercise, and it appears as part of the key results in several applications. Examples 4.2.6 trough 4.2.9 are based on simplifications of examples that can be found in [11].

Example 4.2.6. *Let* $f : \mathbb{R}^n \to \mathbb{R}^m$ *be continuously differentiable and let* $g \in (C[t_0, t_1])^n$. *With these definitions of* f *and* g, *the operator*

$$T : g \to (f \circ g)(t_1) = f(g(t_1))$$
$$T : (C[t_0, t_1])^n \to \mathbb{R}^m$$

is Fréchet differentiable. By definition, we must show that

$$T(x+h) - T(x) - DT(x) \circ h = o\left(\|h\|_{(C[t_0,t_1])^n}\right)$$

for some bounded linear operator $DT(x)$ that maps

$$DT(x) : (C[t_0, t_1])^n \to \mathbb{R}^m.$$

Since f is continuously differentiable on all of $\mathbb{R}^n$, we know that the partial derivatives $\frac{\partial f}{\partial \xi}$ exist and are continuous on all of $\mathbb{R}^n$. In addition, we know that the Taylor series approximation

$$f(\xi + \eta) = f(\xi) + \frac{\partial f}{\partial \xi}(\xi)\eta + o\left(\|\eta\|_{\mathbb{R}^n}\right) \tag{4.5}$$

holds. Suppose $x, h \in (C[t_0, t_1])^n$. Equation (4.5) is evaluated at $\xi = x(t_1) \in \mathbb{R}^n$ and $\eta = h(t_1) \in \mathbb{R}^n$

$$f\left(x(t_1) + h(t_1)\right) = f\left(x(t_1)\right) + \frac{\partial f}{\partial \xi}\left(x(t_1)\right)h(t_1) + o\left(\|h(t_1)\|_{\mathbb{R}^n}\right).$$

By definition, we have

$$T(x + h) - T(x) - \frac{\partial f}{\partial \xi}\left(x(t_1)\right)h(t_1) = o\left(\|h(t_1)\|_{\mathbb{R}^n}\right).$$

In other words,

$$\frac{\|T(x + h) - T(x) - \frac{\partial f}{\partial \xi}\left(x(t_1)\right)h(t_1)\|}{\|h(t_1)\|_{\mathbb{R}^n}} \to 0 \tag{4.6}$$

as

$$\|h(t_1)\|_{\mathbb{R}^n} \to 0. \tag{4.7}$$

This is nearly the result we seek. If we could replace $\|h(t_1)\|_{\mathbb{R}^n}$ with the norm $\|h\|_{(C[t_0,t_1])^n}$ in Equations (4.6) and (4.7), the desired result would follow. However, we know that

$$\|h\|_{(C[t_0,t_1])^n} = \sup_{t \in [t_0, t_1]} \|h(t)\|_{\mathbb{R}^n} \geq \|h(t_1)\|_{\mathbb{R}^n}.$$

Consequently, we have the inequality

$$\frac{\|T(x+h)-T(x)-\frac{\partial f}{\partial \xi}(x(t_1))h(t_1)\|}{\|h(t_1)\|_{\mathbb{R}^n}} \geq \frac{\|T(x+h)-T(x)-\frac{\partial f}{\partial \xi}(x(t_1))h(t_1)\|}{\|h\|_{(C[t_0,t_1])^n}}.$$

We have shown that

$$\frac{\|T(x + h) - T(x) - \frac{\partial f}{\partial \xi}\left(x(t_1)\right)h(t_1)\|}{\|h\|_{(C[t_0,t_1])^n}} \to 0$$

as $\|h\|_{(C[t_0,t_1])^n} \to 0$. Hence, the Fréchet derivative is given by

$$DT(x) \circ h \triangleq \frac{\partial f}{\partial \xi}\left(x(t_1)\right)h(t_1).$$

In the next example, we study the difference in establishing that a functional is Gateaux differentiable, in comparison to Fréchet differentiable. We will see that in some cases, considerably more work is necessary to establish the Fréchet differentiability of a functional. A primary tool in the current example is Theorem 4.1.2.

Example 4.2.7. *Again, suppose $f : \mathbb{R} \times \mathbb{R}^n \to \mathbb{R}^m$ and $f(t, x)$ is continuously differentiable with respect to x. The function $g : [t_0, t_1] \to \mathbb{R}^n$ is assumed to be continuous. Define the shorthand notation for two Cartesian product spaces:*

$$X \triangleq (C[t_0, t_1])^n, \qquad Y \triangleq (C[t_0, t_1])^m,$$

A functional $F : X \to Y$ is defined as

$$(F(g))(t) \triangleq f(t, g(t)).$$

By hypothesis, the function $f(t, x)$ is continuously differentiable with respect to x at all points $x \in \mathbb{R}^n$. Consequently,

$$
\begin{aligned}
\lim_{\epsilon \to 0} & \frac{F(g + \epsilon \eta) - F(g)}{\epsilon} \\
&= \lim_{\epsilon \to 0} \frac{f(t, g(t) + \epsilon \eta(t)) - f(t, g(t))}{\epsilon} \\
&= \lim_{\epsilon \to 0} \frac{1}{\epsilon} \left\{ f(t, g(t)) + \frac{\partial f}{\partial x}(t, g(t)) \, \epsilon \eta(t) + o(\|\epsilon \eta(t)\|) - f(t, g(t)) \right\} \\
&= \frac{\partial f}{\partial x}(t, g(t)) \, \eta(t) \\
&= D_G F(g) \circ \eta.
\end{aligned}
$$

Since f is globally continuously differentiable, $D_G F(g)(\eta)$ exists for any g or η in X. Obviously,

$$D_G F(g) \in \mathcal{L}(X, Y).$$

We have

$$
\begin{aligned}
\|D_G F(g) \circ \eta\|_Y &= \sup_{t \in [t_0, t_1]} \| \tfrac{\partial f}{\partial x}(t, g(t)) \, \eta(t) \|_{\mathbb{R}^m} \\
&\leq \sup_{t \in [t_0, t_1]} \left(\| \tfrac{\partial f}{\partial x}(t, g(t)) \|_{\mathbb{R}^{m \times n}} \|\eta(t)\|_{\mathbb{R}^n} \right) \\
&\leq \mathcal{C}(g) \sup_{t \in [t_0, t_1]} \|\eta(t)\|_{\mathbb{R}^n} \\
&= \mathcal{C}(g) \|\eta\|_X
\end{aligned}
$$

where

$$\mathcal{C}(g) \triangleq \sup_{t \in [t_0, t_1]} \left\| \frac{\partial f}{\partial x}(t, g(t)) \right\|_{\mathbb{R}^{m \times n}}.$$

The constant $\mathcal{C}(g)$ is finite for each $g \in X$. This follows since the entries of the Jacobian $\frac{\partial f}{\partial x}$ are continuous. The composition of continuous functions is continuous and every continuous function defined on a compact set attains its maximum. Finally, we claim that the map

$$g \to D_G F(g)$$

is continuous from X into $\mathcal{L}(X,Y)$. By definition,

$$
\begin{aligned}
\|D_G F(\xi) - D_G F(\gamma)\| &= \sup_{\eta \neq 0} \frac{\|(D_G F(\xi) - D_G F(\gamma)) \circ \eta\|_Y}{\|\eta\|_X} \\
&= \sup_{\eta \neq 0} \frac{\left\{ \sup_{t \in [t_0,t_1]} \left\| \left(\frac{\partial f}{\partial x}(t,\xi(t)) - \frac{\partial f}{\partial x}(t,\gamma(t)) \right) \eta(t) \right\|_{\mathbb{R}^m} \right\}}{\|\eta\|_X} \\
&\leq \sup_{\eta \neq 0} \frac{\left\{ \sup_{t \in [t_0,t_1]} \left\| \frac{\partial f}{\partial x}(t,\xi(t)) - \frac{\partial f}{\partial x}(t,\gamma(t)) \right\|_{m,n} \|\eta(t)\|_{\mathbb{R}^n} \right\}}{\|\eta\|_X} \\
&\leq \sup_{\eta \neq 0} \frac{\left\{ \sup_{t \in [t_0,t_1]} \left\| \frac{\partial f}{\partial x}(t,\xi(t)) - \frac{\partial f}{\partial x}(t,\gamma(t)) \right\|_{m,n} \|\eta\|_X \right\}}{\|\eta\|_X} .
\end{aligned}
$$

Recall that the notation $\| \cdot \|_{m,n}$ denotes the matrix norm. Suppose $\xi \to \gamma$ in X. Then we have

$$
\sup_{t \in [t_0,t_1]} \left\| \frac{\partial f}{\partial x}(t,\xi(t)) - \frac{\partial f}{\partial x}(t,\gamma(t)) \right\|_{m,n} \to 0 \tag{4.8}
$$

since the function $\frac{\partial f}{\partial x}(t,x)$ is uniformly continuous on some compact set Ω chosen large enough so that $\xi(t), \gamma(t) \in \Omega \quad \forall\, t \in [t_0,t_1]$. The convergence shown in Equation (4.8) follows from the following proposition

Proposition 4.2.1. *Suppose $f : \mathbb{R} \to \mathbb{R}$ is uniformly continuous and $\gamma_k \to \gamma$ in $C[t_1,t_2]$. Then*

$$
f \circ \gamma_k \to f \circ \gamma \quad in \quad C[t_1,t_2].
$$

Proof. We must show that for any $\epsilon > 0$, $\exists\, k_0 \in \mathbb{N}$ such that

$$
k \geq k_0 \quad \Longrightarrow \quad \left| (f \circ \gamma_k)(t) - (f \circ \gamma)(t) \right| < \epsilon \quad \forall\, t \in [t_0,t_1].
$$

Fix $\epsilon > 0$. By the definition of uniform continuity, there is a $\delta > 0$ such that

$$
|\xi - \eta| < \delta \quad \Longrightarrow \quad \left| f(\xi) - f(\eta) \right| < \epsilon \quad \forall\, \xi, \eta \in \mathbb{R}.
$$

Since $\gamma_k \to \gamma$ in $C[t_1,t_2]$, for each $\tilde{\epsilon} > 0$ there is a $k_0 \in \mathbb{N}$ such that

$$
k \geq k_0 \quad \Longrightarrow \quad \left| \gamma_k(t) - \gamma(t) \right| < \tilde{\epsilon} \quad \forall\, t \in [t_0,t_1].
$$

In particular, for the choices $\tilde{\epsilon} = \delta$, there is a $k_0 \in \mathbb{N}$ such that

$$
k \geq k_0 \quad \Longrightarrow \quad \left| \gamma_k(t) - \gamma(t) \right| < \delta \quad \forall\, t \in [t_0,t_1].
$$

But this means that

$$
k \geq k_0 \quad \Longrightarrow \quad \left| f\big(\gamma_k(t)\big) - f\big(\gamma(t)\big) \right| < \epsilon \quad \forall\, t \in [t_0,t_1]. \qquad \square
$$

By Corollary (4.1.1), the functional F is in fact Fréchet differentiable and $D_F F = D_G F$.

Example 4.2.8. *This example is but a slight generalization of Example 4.2.7. Suppose that* $f : \mathbb{R}^n \times \mathbb{R}^m \times \mathbb{R} \to \mathbb{R}^p$

$$f(x, u, t) = \big(f_1(x, u, t), f_2(x, u, t), \ldots, f_p(x, u, t)\big).$$

Each of the functions $f_i : \mathbb{R}^n \times \mathbb{R}^m \times \mathbb{R} \to \mathbb{R}$ *is assumed to be continuously differentiable with respect to* x *and* u. *In the remainder of this example, define the Cartesian product spaces*

$$X^0 = \big(C^0[t_0, t_1]\big)^n, \quad X^1 = \big(C^1[t_0, t_1]\big)^n, \quad Y = (C[t_0, t_1])^m, \quad Z = (C[t_0, t_1])^p.$$

We define the functional

$$F : X^1 \times Y \to Z$$
$$[F(x, u)](t) = f\big(x(t), u(t), t\big).$$

Note that, in comparison to the last example, a higher degree of smoothness is assumed for the first argument. We claim that F *is Fréchet differentiable with respect to* x *and* u *and*

$$\Big[(DF(x, u)) \circ (y, v)\Big](t) = \frac{\partial f}{\partial x}\big(x(t), u(t), t\big)y(t) + \frac{\partial f}{\partial u}\big(x(t), u(t), t\big)v(t).$$

Since f *is continuously differentiable with respect to* x *and* u *at all points in* $\mathbb{R}^n \times \mathbb{R}^m$, *we have*

$$\lim_{\epsilon \to 0} \frac{F(x + \epsilon y, u + \epsilon v) - F(y, u)}{\epsilon}$$

$$= \lim_{\epsilon \to 0} \frac{f\big(x(t) + \epsilon y(t), u(t) + \epsilon v(t), t\big) - f\big(x(t), u(t), t\big)}{\epsilon}$$

$$= \lim_{\epsilon \to 0} \frac{1}{\epsilon}\bigg\{ \frac{\partial f}{\partial x}\big(x(t), u(t), t\big)\epsilon y(t) + \frac{\partial f}{\partial u}\big(x(t), u(t), t\big)\epsilon v(t)$$

$$+ o\left(\big\|\epsilon y(t)\big\|_{\mathbb{R}^n} + \big\|\epsilon v(t)\big\|_{\mathbb{R}^m}\right)\bigg\}$$

$$= \frac{\partial f}{\partial x}\big(x(t), u(t), t\big)y(t) + \frac{\partial f}{\partial u}\big(x(t), u(t), t\big)v(t)$$

$$= \Big[(DF(x, u)) \circ (y, v)\Big](t).$$

By definition, we must have

$$DF(x, u) \circ (\cdot, \cdot) : X^1 \times Y \to Z$$

is a bounded and linear operator. It is clear that the operator

$$(y, v) \to DF(x, u) \circ (y, v)$$

is linear. It is likewise bounded since

$$\sup_{t\in[t_0,t_1]} \left\| \frac{\partial f}{\partial x}\big(x(t),u(t),t\big)y(t) + \frac{\partial f}{\partial u}\big(x(t),u(t),t\big)u(t) \right\|_{\mathbb{R}^p}$$

$$= \| DF(u,v) \circ (y,v) \|_Z$$

$$\leq \sup_{t\in[t_0,t_1]} \left(\left\| \frac{\partial f}{\partial x}\big(x(t),u(t),t\big) \right\|_{\mathbb{R}^{p\times n}} \|y(t)\|_{\mathbb{R}^n} \right)$$

$$+ \sup_{t\in[t_0,t_1]} \left(\left\| \frac{\partial f}{\partial u}\big(x(t),u(t),t\big) \right\|_{\mathbb{R}^{p\times m}} \|v(t)\|_{\mathbb{R}^m} \right)$$

$$\leq C_1 \sup_{t\in[t_0,t_1]} \|y(t)\|_{\mathbb{R}^n} + C_2 \sup_{t\in[t_0,t_1]} \|v(t)\|_{\mathbb{R}^m}$$

$$\leq C\left(\|y\|_{X^0} + \|v\|_Y \right)$$

$$\leq C\left(\|y\|_{X^1} + \|v\|_Y \right)$$

where

$$C_1 = \sup_{t\in[t_0,t_1]} \left\| \frac{\partial f}{\partial x}\big(x(t),u(t),t\big) \right\|_{\mathbb{R}^{p\times n}}$$

$$C_2 = \sup_{t\in[t_0,t_1]} \left\| \frac{\partial f}{\partial u}\big(x(t),u(t),t\big) \right\|_{\mathbb{R}^{p\times m}}$$

and

$$C = \max\{C_1, C_2\}.$$

We have yet to show that F is Fréchet differentiable. As in Example 4.2.7, we can argue that the map

$$(x,u) \mapsto DF(x,u)$$

is continuous from

$$X^1 \times Y$$

to

$$\mathcal{L}X^1 \times Y, Z.$$

The desired result then follows. The functional F is Fréchet differentiable by Corollary 4.1.1.

In the last example, it may not have been clear why we required the extra degree of regularity in the first argument of the functional F. In this example, we demonstrate how several of the previous examples provide a framework for studying the differentiability of functionals that contain ordinary derivatives. This is not the only technique, but provides an excellent starting point.

Example 4.2.9. *Let* $g : \mathbb{R}^n \times \mathbb{R}^m \times \mathbb{R} \to \mathbb{R}^n$ *and suppose the function*

$$(x, u) \to g(x, u, t)$$

is continuously differentiable. Define the Cartesian product spaces X^1, X^0, Y *as in Example 4.2.8. Define the mapping*

$$G : X^1 \times Y \to X^0$$
$$G(x, u)(t) \triangleq \dot{x}(t) - g\big(x(t), u(t), t\big).$$

Define

$$\mathcal{X} = X^1 \times Y$$
$$\mathcal{Y} = X^0.$$

We claim that the function G *is Fréchet differentiable and that*

$$\Big[DG(x, u) \circ (y, v)\Big](t) = \dot{y}(t) - \frac{\partial g}{\partial x}\big(x(t), u(t), t\big)y(t) - \frac{\partial g}{\partial u}\big(x(t), u(t), t\big)v(t).$$

The function G *is the composition of two mappings*

$$G = G_1 - G_2$$

where

$$\big[G_1(x, u)\big](t) = \dot{x}(t)$$
$$\big[G_2(x, u)\big](t) = g\big(x(t), u(t), t\big).$$

Each of these mappings are, in fact, Fréchet differentiable. The map

$$(y, v) \to \dot{y}$$

from X *to* Y *is linear and bounded. This fact is clear since*

$$\sup_{t \in [t_0, t_1]} \|\dot{y}(t)\|_{\mathbb{R}^n} = \|\dot{y}\|_{X^0}$$
$$\leq \|y\|_{X^1}$$
$$\leq \|y\|_{X^1} + \|v\|_Y$$
$$= \|(y, v)\|_{\mathcal{X}}.$$

We have

$$\left|\big[G_1(x + y, u + v) - G_1(x, u)\big](t) - \dot{y}(t)\right| \equiv 0$$

and consequently

$$\lim_{\|(y,v)\|_{\mathcal{X}} \to 0} \frac{\big\|G_1(x + y, u + v) - G_1(x, u) - \dot{y}\big\|_Y}{\|(y, v)\|_{\mathcal{X}}} = 0.$$

We have shown, thereby, that

$$DG_1(x, u) \circ (y, v) = \dot{y}.$$

Moreover, by Example 4.2.8, G_2 is Fréchet differentiable and

$$[DG_2(x, u) \circ (y, v)](t) = \frac{\partial g}{\partial x}(x(t), u(t), t)y(t) + \frac{\partial g}{\partial u}(x(t), u(t), t)v(t).$$

The proof is complete.

The last several examples have been associated with the Banach spaces of continuous functions, $C^m(\Omega)$. Functionals on these spaces play an important role in the optimization of systems governed by ordinary differential equations. We now turn to examples that focus on types of quadratic functionals. These functionals appear in many control and identification problems.

Example 4.2.10. *In this example we study the Gateaux differential of the quadratic functional*

$$J(w) = \int_0^T \|Cw(t) - \tilde{w}(t)\|_{\mathbb{R}^n}^2 \, dt$$

where $C \in \mathbb{R}^{n \times m}$, $w \in L^2((0, T), \mathbb{R}^m)$, and $\tilde{w}(t) \in L^2((0, T), \mathbb{R}^n)$. There is simply an enormous collection of control and mechanics problems in which we seek to minimize quadratic functionals. Quadratic functionals are perhaps the simplest nonnegative functionals that admit differentiability properties. By definition, we can write

$$
\begin{aligned}
DJ(w, v) &= \lim_{\epsilon \to 0} \frac{1}{\epsilon} \left\{ \int_0^T \|C((w(t) + \epsilon v(t)) - \tilde{w}(t)\|_{\mathbb{R}^n}^2 \right. \\
&\quad \left. - \int_0^T \|Cw(t) - \tilde{w}(t)\|_{\mathbb{R}^n}^2 \, dt \right\} \\
&= \lim_{\epsilon \to 0} \frac{1}{\epsilon} \left\{ \int_0^T \|Cw(t) - \tilde{w}(t) + \epsilon Cv(t)\|_{\mathbb{R}^n}^2 \right. \\
&\quad \left. - \int_0^T \|Cw(t) - \tilde{w}(t)\|_{\mathbb{R}^n}^2 \, dt \right\} \\
&= \lim_{\epsilon \to 0} \frac{1}{\epsilon} \left\{ 2\epsilon \int_0^T (Cw(t) - \tilde{w}(t), Cv(t))_{\mathbb{R}^n} \, dt \right. \\
&\quad \left. + \epsilon^2 \int_0^T (Cv(t), Cv(t))_{\mathbb{R}^n} \, dt \right\} \\
&= 2 \int_0^T (Cw(t) - \tilde{w}(t), Cv(t))_{\mathbb{R}^n} \, dt.
\end{aligned}
$$

As in the previous cases, it remains to show that $DJ(w, \cdot)$ is linear and bounded. It is trivial to verify that

$$DJ(w, v + \alpha u) = DJ(w, v) + \alpha DJ(w, u).$$

Furthermore, by the Cauchy-Schwarz inequality, we have

$$
\begin{aligned}
|DJ(w,v)| &= 2\left| \int_0^T \Big(Cw(t) - \tilde{w}(t), Cv(t) \Big)_{\mathbb{R}^n} dt \right| \\
&\leq 2 \int_0^T \|Cw(t) - \tilde{w}(t)\|_{\mathbb{R}^n} \|Cv(t)\|_{\mathbb{R}^n}\, dt \\
&\leq 2 \|Cw - \tilde{w}\|_{L^2((0,T),\mathbb{R}^n)} \|C\|_{\mathbb{R}^{n\times m}} \|v\|_{L^2((0,T),\mathbb{R}^m)} \\
&\leq \mathcal{C}\|v\|_{L^2((0,T),\mathbb{R}^m)}.
\end{aligned}
$$

We have consequently shown that

$$
DJ(w) \circ v = 2 \int_0^T (Cw(t) - \tilde{w}(t), Cv(t))_{\mathbb{R}^n}\, dt.
$$

The last example was cast in very conventional terms. The functions took values in $\mathbb{R}^n$. The inner products and Hilbert spaces were all the most familiar that are encountered in applications. In the next example, we consider a natural extension of Example 4.2.11 to more abstract spaces.

Example 4.2.11. *In this example, we show how the calculation of the Gateaux derivative of a class of functionals defined over time-varying vectors*

$$
w \in L^2((0,T), \mathbb{R}^m)
$$

can be extended to quadratic functionals defined on general Hilbert spaces. Let H be a real Hilbert space with inner product $(\cdot,\cdot)_H$ and suppose the $C \in \mathcal{L}(X,H)$ where X is a normed vector space. We now consider the calculation of the Gateaux differential of the quadratic functional

$$
J(w) = \int_0^T \|Cw(t) - \tilde{w}(t)\|_H^2\, dt
$$

where $w \in L^2((0,T), X)$ and $\tilde{w} \in L^2((0,T), H)$. First, note that since $w \in L^2((0,T), X)$, we know that

$$
Cw \in L^2((0,T), H)
$$

since

$$
\int_0^T \|Cw(t)\|_H^2\, dt \leq \|C\|_{\mathcal{L}(X,H)}^2 \int_0^T \|w(t)\|_X^2\, dt.
$$

Thus, the cost functional makes sense for any $w \in L^2((0,T), X)$. Following the line of attack taken in the last example, we can directly calculate

$$
\begin{aligned}
DJ(w, v) &= \lim_{\epsilon \to 0} \frac{1}{\epsilon} \left\{ \int_0^T \|C\left(w(t) + \epsilon v(t)\right) - \tilde{w}(t)\|_H^2 dt \right. \\
&\quad \left. - \int_0^T \|Cw(t) - \tilde{w}(t)\|_H^2 dt \right\} \\
&\;\;\vdots \\
&= \lim_{\epsilon \to 0} \frac{1}{\epsilon} \left\{ 2\epsilon \int_0^T \left(Cw(t) - \tilde{w}(t), Cv(t) \right)_H dt + O(\epsilon^2) \right\} \\
&= 2 \int_0^T \left(Cw(t) - \tilde{w}(t), Cv(t) \right)_H dt.
\end{aligned}
$$

Again, it is trivial to prove that

$$
DJ(w, v + \alpha u) = DJ(w, v) + \alpha DJ(w, u)
$$

and hence $DJ(w, \cdot)$ is linear. The mapping

$$
v \to DJ(w, v)
$$

is also bounded since

$$
\left| \int_0^T \left(Cw(t) - \tilde{w}(t), Cv(t) \right)_H dt \right| \leq \|Cw - \tilde{w}\|_{L^2((0,T),H)} \|Cv\|_{L^2((0,T),H)}
$$

$$
\leq \mathcal{C} \|v\|_{L^2((0,T),X)}.
$$

As in the last example, we have shown that the Gateaux derivative can be written as

$$
DJ(w) \circ v = 2 \int_0^T \left(Cw(t) - \tilde{w}(t), Cv(t) \right)_H dt. \tag{4.9}
$$

In this last example in this section, we study what is perhaps, the most well-known example of the use of the calculus of variations. The result of the derivation of the functional studied below is the Euler-Lagrange equations.

Example 4.2.12. *Perhaps the most well-known example of an integral functional that has been studied in variational calculus is the Lagrange functional*

$$
L(y) \equiv \int_{t_0}^{t_1} f(y(t), \dot{y}(t), t) dt
$$

where the given function f is continuous and has continuous first partial derivatives in each of its arguments. Define the domain of L to be a subset of $C^1[t_0, t_1]$ that contains functions that assume two given values y_0 and y_1 at the endpoints of the interval $[t_0, t_1]$. That is, the domain of the functional L is given by

$$\mathcal{D}(L) = \{y \in C^1[t_0, t_1] : y(t_0) = y_0 \text{ and } y(t_1) = y_1\}.$$

We must derive the Gateaux differential of L at $y \in \mathcal{D}(L)$ in the z direction. To begin with, we note several qualitative properties of the functional L defined on the set $\mathcal{D}(L)$. Because $\mathcal{D}(L) \subseteq C^1[t_0, t_1]$, $\dot{y}(t)$ exists for all $y \in \mathcal{D}(L)$, and because the composition of continuous functions is continuous, the integrand is a continuous function over the closed and bounded interval $[t_0, t_1]$. Consequently, $L(y)$ is finite for any $y \in \mathcal{D}(L)$. Finally, it is an elementary exercise to show that $\mathcal{D}(L)$ is in fact a convex set in $C^1[t_0, t_1]$. In contrast to many of the preceding examples, however, the domain of L is not a subspace. Given any two elements

$$y, z \in \mathcal{D}(L)$$

it is not true that $\alpha y + \beta z \in \mathcal{D}(L)$ for any choice of $\alpha, \beta \in \mathbb{R}$. Indeed, for $\alpha y + \beta z \in C^1[a, b]$ with arbitrary real α and β, we have

$$(\alpha y + \beta z)|_{t_0} = (\alpha + \beta)y_0 \neq y_0.$$

The boundary conditions are not satisfied by sums of elements extracted from $\mathcal{D}(L)$. The Gateaux differential can be calculated along certain directions, however. By the definition of the Gateaux differential, we can write

$$DL(y, z) = \lim_{\epsilon \to 0} \frac{1}{\epsilon} \left\{ \int_{t_0}^{t_1} f(y(t) + \epsilon z(t), \dot{y}(t) + \epsilon \dot{z}(t), t)dt \right.$$
$$\left. - \int_{t_0}^{t_1} f(y(t), \dot{y}(t), t)dt \right\}.$$

Because L maps into the real line, $\epsilon \mapsto L(y + \epsilon z)$ is a real-valued function of ϵ. If, in addition, the map $\epsilon \mapsto L(y + \epsilon z)$ is classically differentiable, we can write

$$DL(y, z) = \lim_{\epsilon \to 0} \frac{L(y + \epsilon z) - L(y)}{\epsilon}$$
$$= \frac{dL}{d\epsilon}(y + \epsilon z)\bigg|_{\epsilon=0}$$
$$= \left(\lim_{\delta \to 0} \frac{L(y + (\epsilon + \delta)z - L(y + \epsilon z))}{\delta} \right)\bigg|_{\epsilon=0}.$$

This identity can be quite useful when calculating the Gateaux differential of real-valued functionals. For the case at hand, L is an integral functional. By Leibniz's rule, we can write

$$DL(y,z) = \frac{dL}{d\epsilon}(y + \epsilon z)\Big|_{\epsilon=0}$$

$$= \frac{d}{d\epsilon}\left\{\int_{t_0}^{t_1} f(y(t) + \epsilon z(t), \dot{y}(t) + \epsilon \dot{z}(t), t)dt\right\}\Big|_{\epsilon=0}$$

$$= \left\{\int_{t_0}^{t_1} \frac{d}{d\epsilon}\left\{f(y(t) + \epsilon z(t), \dot{y}(t) + \epsilon \dot{z}(t), t)\right\}dt\right\}\Big|_{\epsilon=0}.$$

By differentiating the integrand and using the chain rule, we obtain

$$DL(y,z) = \int_{t_0}^{t_1}\left\{\frac{\partial f}{\partial y}(y + \epsilon z, \dot{y} + \epsilon \dot{z}, t)z(t)\right.$$

$$\left. + \frac{\partial f}{\partial \dot{y}}(y + \epsilon z, \dot{y} + \epsilon \dot{z}, t)\dot{z}(t)\right\}dt\Big|_{\epsilon=0}. \qquad (4.10)$$

Evaluating the expression at $\epsilon = 0$, we obtain

$$DL(y,z) = \int_{t_0}^{t_1}\left\{\frac{\partial f}{\partial y}\big(y(t), \dot{y}(t), t\big)z(t) + \frac{\partial f}{\partial \dot{y}}\big(y(t), \dot{y}(t), t\big)\dot{z}(t)\right\}dt.$$

If, in addition, the mapping

$$t \mapsto \frac{\partial f}{\partial \dot{y}}\big(y(t), \dot{y}(t), t\big) \qquad (4.11)$$

is classically differentiable, we can integrate by parts to achieve

$$DL(y,z) = \int_{t_0}^{t_1}\left\{\frac{\partial f}{\partial y}(y(t), \dot{y}(t), t) - \frac{d}{dt}\frac{\partial f}{\partial \dot{y}}(y(t), \dot{y}(t), t)\right\}z(t)dt$$

$$+ \left(\frac{\partial f}{\partial \dot{y}}(y(t), \dot{y}(t), t)z(t)\right)\Big|_{t_0}^{t_1}.$$

Since we require that $y + \epsilon z \in \mathcal{D}(L)$ for all ϵ small enough, and $y \in \mathcal{D}(L)$, we require that $z(t_0) = 0$ and $z(t_1) = 0$. The above expression consequently reduces to

$$DL(y,z) = \int_{t_0}^{t_1}\left\{\frac{\partial f}{\partial y}(y, \dot{y}, t) - \frac{d}{dt}\left(\frac{\partial f}{\partial \dot{y}}(y, \dot{y}, t)\right)\right\}z(t)dt.$$

Again, we emphasize that we have calculated the Gateaux differential in a particular direction. We have chosen

$$z \in \left\{C^1[t_0, t_1]\colon z(t_0) = z(t_1) = 0\right\}.$$

In reviewing this development, it may seem awkward to hypothesize the differentiability of the map appearing in Equation (4.11). However, as shown in the following theorem, this is often not a serious restriction when we seek to characterize extrema of functionals.

Theorem 4.2.1. *Suppose that*

$$DL(y, z) = 0 \quad \forall z \in C^1[t_0, t_1] \cap \{z : z(t_0) = z(t_1) = 0\}$$

where $DL(y, z)$ is given in Equation (4.10). Then the mapping

$$t \mapsto \frac{\partial f}{\partial \dot{y}}(y(t), \dot{y}(t), t)$$

is differentiable.

The interested reader is referred to Proposition 4.2 in [32, p. 98] or Section 7.5 in [17, pp. 179–181] for a proof.

Chapter 5

Minimization of Functionals

The collection of intertwining definitions, theorems, lemmas and propositions that constitute optimization theory in a functional analytic setting can be intimidating for those who are not full-time mathematicians. However, the task of utilizing the functional analytic framework in practical problems is made tractable by keeping in mind that all of these concepts are generalizations of much simpler, more familiar ideas. In this chapter we will emphasize that an understanding of minimization of convex functionals over abstract spaces is facilitated by drawing analogies to some well-known facts from elementary calculus. At the heart of any optimization problem, whether it arises in structural design or control design, it is necessary to find the extrema of some function that represents a cost, usually subject to a set of constraints. To solve this problem, at least three "natural" questions arise:

- Is the optimization problem at least well posed? That is, does there exist an optimal solution for the problem as stated?
- How can we find the extrema that represent a solution of the optimization problem?
- Can we guarantee that the extrema we have found are the minima or maxima we sought?

In the language of mathematics, the first question above is a statement of the well-posedness and existence of a solution to an optimization problem, while the second question seeks to characterize the optimal solutions.

5.1 The Weierstrass Theorem

If the measure of the utility of any theorem is judged by how concisely it may be expressed, and how widely it may be applied, then the Weierstrass Theorem rightly plays a central role in optimization theory. It provides sufficient conditions for the solution of the optimization problem where we seek to find $u \in \mathcal{U} \subseteq X$ such that

$$f(u) = \inf_{v \in \mathcal{U}} f(v)$$

without requiring the differentiability properties for f, stated in Theorem 5.3.2. It is important to note that this theorem combines *continuity* and *compactness* requirements to guarantee the solution of the optimization problem.

Theorem 5.1.1. *Let (X, τ) be a compact topological space and let the functional $f : X \to \overline{\mathbb{R}}$ be continuous. Then there exists an $x_0 \in X$ such that*

$$f(x_0) = \inf_{x \in X} f(x).$$

Because the proof of this theorem, while well-known, is instructive and serves as a model for the proofs of more general results in this chapter, we will summarize it here. We will require the following alternative characterizations of continuity on topological spaces to carry out this proof in a manner that can be "lifted" to more general circumstances. Recall that one of the most common definitions of continuity is cast in terms of inverse images of open sets.

Definition 5.1.1. *Let (X, τ_x) and (Y, τ_y) be topological spaces. A function $f : X \to Y$ is continuous at $x_0 \in X$ if the inverse image of every open set $\mathcal{O}$ in Y that contains $f(x_0)$ is an open set in X that contains x_0. That is*

$$\mathcal{O} \in \tau_y \text{ and } f(x_0) \in \mathcal{O} \implies x_0 \in f^{-1}(\mathcal{O}) \in \tau_x.$$

A function that is continuous at each point of a topological space is said to be continuous on that space. The following two definitions restate this fact.

Definition 5.1.2. *Let (X, τ_x) and (Y, τ_y) be topological spaces. The following are equivalent:*

(i) *$f : X \to Y$ is continuous.*
(ii) *The inverse image under f of every open set in Y is an open set in X. That is,*

$$\mathcal{O} \in \tau_y \to f^{-1}(\mathcal{O}) \in \tau_x.$$

(iii) *The inverse image of every closed set in Y is a closed set in X.*

Now we return to the proof of Theorem 5.1.1.

Proof. Suppose that $\alpha = \inf_{x \in X} f(x)$. Consider the sequence of closed sets in the range of f

$$Q_k = \left[\alpha - \frac{1}{k}, \alpha + \frac{1}{k}\right] \qquad k \in \mathbb{N}.$$

From this sequence of closed sets, we can construct a sequence of closed sets in the domain X

$$C_k = \{x \in X : f(x) \in Q_k\}.$$

By construction, this sequence of sets is nested

$$C_{k+1} \subseteq C_k \quad \forall k \in \mathbb{N}$$

and each C_k is compact being a closed subset of a compact set. The sequence of compact sets $\{C_k\}_{k=1}^{\infty}$ clearly satisfies the finite intersection property, so that

$$\exists\, x_0 \in \bigcap_{k=1}^{\infty} C_k.$$

It consequently follows that

$$f(x_0) = \alpha = \inf_{x \in X} f(x). \qquad \qquad \Box$$

5.2 Elementary Calculus

To set the foundation for the analysis that follows, let us review some well-known results from elementary calculus that give relatively straightforward methods for answering these questions for cost functions defined in terms of a single real variable. When we express an optimization problem in terms of a function f that maps the real line into itself, that is,

$$f : \mathbb{R} \to \mathbb{R} \tag{5.1}$$

we seek to find a real number $x_0 \in \mathbb{R}$ such that

$$f(x_0) = \inf_{x \in C} f(x) \tag{5.2}$$

where C is the constraint set. Now, there are many ways in which we can construct simple functions for which there is no minimizer over the constraint set. If the constraint set is *unbounded*, such as the entire real line, an increasing function like $f(x) = x$ obviously does not have a minimizer. Even if the constraint set is bounded, for example $C \equiv (0, 1]$, there is no minimizer for the simple function $f(x) = x$. Intuitively, we would like to say that $x_0 = 0$ is the minimizer, but this point is not in the constraint set C. While there are many theorems that can describe when a function will achieve its minimum over some constraint set, one prototypical example is due to Weierstrass.

Theorem 5.2.1. *If f is a real-valued function defined on a closed and bounded subset C of the real line, then f achieves its minimum.*

If f is in fact a differentiable function of the real variable x, and is defined on all of $\mathbb{R}$, then the problem of characterizing the values of x where extrema may occur is well known: the extrema may occur only when the derivative of the function f vanishes. From elementary calculus we know that:

Theorem 5.2.2. *If f is a differentiable, real-valued function of the real variable x and is defined on all of $\mathbb{R}$, then*

$$f(x_0) = \inf_{x \in \mathbb{R}} f(x) \qquad \textit{implies that} \qquad f'(x_0) = 0.$$

In fact, most students studying calculus for the first time spend a great deal of time finding the zeros of the derivative of a function, in order to find the extrema of the function. Soon after learning that the first derivative can be used to characterize the possible locations of the extrema of a real-valued function, the student of calculus is taught to examine the second derivative of a function to gain some insight into the nature of the extrema.

Theorem 5.2.3. *If f is a twice differentiable, real-valued function defined on all of $\mathbb{R}$, $f'(x_0) = 0$, and*

$$f''(x_0) > 0 \qquad\qquad (5.3)$$

then x_0 is a relative minimum. In other words,

$$f(x) \geq f(x_0)$$

for all x in some neighborhood of x_0.

Of course, readers will recognize these theorems immediately. These theorems are fundamentals of the foundations of real variable calculus, and require no abstract, functional analytic framework whatsoever. Because of their simple form and graphical interpretation, they are easy to remember. They are important to this chapter in that they provide a touchstone for more abstract results in functional analysis that are required to treat optimization problems in mechanics. For example, the elastic energy stored in a beam, rod, plate, or membrane cannot be expressed in terms of a real-valued function of a real variable $f(x)$. One can hypothesize that equilibria of these structures correspond to minima in their stored energy, but the expressions for the stored energy are not *classically* differentiable functions of a real variable. We cannot simply differentiate the energy expressions in a classical sense to find the equilibria as described in the above theorems.

What is required, then, is a generalization of these theorems that is sufficiently rich to treat the meaningful collection of problems in mechanics and control theory. For a large class of problems, we will find that each of the simple, intuitive theorems above can be generalized so that they are meaningful for problems in control and mechanics. In particular, this chapter will show that:

- The Weierstrass Theorem can be generalized to a functional analytic framework. To pass to the treatment of control and mechanics problems, we will need to generalize the idea of considering *closed and bounded subsets* of the real line, and consider *compact subsets* of topological spaces. We will need to generalize the notion of *continuity of functions* of a real variable to *continuity of functionals* on topological spaces.
- The characterization of minima of real-valued functions by derivatives that vanish will be generalized by considering Gateaux and Fréchet derivatives of functionals on abstract spaces. It will be shown that Theorem 5.2.1 has an immediate generalization to a functional analytic setting.
- The method of determining that a given extrema of a real-valued function is a relative minima, by checking to see if its second derivative is positive, also has a simple generalization. In this case, a relative minima can be deduced if the second Gateaux derivative is positive.

5.3   Minimization of Differentiable Functionals

Now we can state our first step in "lifting" the results from elementary calculus for characterizing minima of a real-valued function, described in Equations (5.1)–(5.3). We consider only functionals having relatively strong differentiability properties to begin, and weaken these assumptions in subsequent sections. It is important to note that the results in this section are *strictly local* in character. That is, if X is a normed vector space and $f : X \to \overline{\mathbb{R}}$ is an extended functional, f is said to have *a local minima* at $x_0 \in X$ if there exists a neighborhood $\mathcal{N}(x_0)$ such that

$$f(x_0) \leq f(y) \quad \forall\, y \in \mathcal{N}(x_0).$$

This is clearly a result that is directly analogous to the local character of the characterization of extrema of real-valued functions. In fact, the primary results of this section are derived by exploiting the identification of $f(x_0 + th)$ with a real-valued function

$$g(t) \equiv f(x_0 + th)$$

where $t \in [0, 1]$ and $h \in X$. Note that for fixed $x_0, h \in X$, $g(t)$ is a real-valued function. Indeed, if g is sufficiently smooth, uniformly for all x_0 and h in some subset of X, we can expand g in a Taylor series about $t = 0$

$$g(t) = g(0) + \sum_{k=1}^{n} \frac{t^k g^{(k)}(0)}{k!} + R_{n+1}.$$

Now we obtain the most direct, simple generalization of Theorem 5.2.2 for real-variable functions.

Theorem 5.3.1. *Let X be a normed vector space, and let $f : X \to \overline{\mathbb{R}}$. If f has a local minimum at $x_0 \in X$ and the Gateaux derivative $Df(x_0)$ exists, we have*

$$\langle Df(x_0), h \rangle_{X^* \times X} = 0 \quad \forall\, h \in X.$$

Proof. By assumption, the limit

$$\langle Df(x_0), h \rangle_{X^* \times X} = \lim_{t \to 0} \frac{f(x_0 + th) - f(x_0)}{t}$$

exists. Since x_0 is a local minimum, we have that

$$\frac{f(x_0 + th) - f(x_0)}{t} \geq 0 \quad \forall\, x_0 + th \in \mathcal{N}(x_0).$$

But as we take the limit as $t \to 0$ for $t > 0$, we always have $x_0 + th \in \mathcal{N}(x_0)$ for t small enough, for any $h \in X$. This fact implies that

$$\langle Df(x_0), h \rangle_{X^* \times X} \geq 0 \quad \forall\, h \in X.$$

By choosing $h = \pm\xi$, we can write

$$\pm \langle Df(x_0), \xi \rangle_{X^* \times X} \geq 0$$

and consequently we obtain $Df(x_0) \equiv 0 \in X^*$.   □

For some functionals that have higher order differentiability properties, it is also possible to express *sufficient* conditions for the existence of local extrema in terms of *positivity* of the differentials. These conditions appear remarkably similar to results from the calculus of real-variable functions.

Theorem 5.3.2. *Let X be a normed vector space and let the functional $f : X \to \overline{\mathbb{R}}$. Suppose that n is an even number with $n \geq 2$ and*
  (i) *f is n times Fréchet differentiable in a neighborhood of x_0,*
  (ii) *$Df^{(n)}$ is continuous at x_0, and*
  (iii) *the nth derivative is coercive, that is*

$$\langle Df^{(n)}(x_0), (h, h, \ldots, h) \rangle_{X^* \times X} \geq c\|h\|_X^n.$$

Then f has a strict local minimum at x_0.

Proof. The proof is left as an exercise.   □

5.4   Equality Constrained Smooth Functionals

In the last section, we have discussed necessary conditions for existence of a local minimum for unconstrained optimization problems. We now consider constrained optimization problems where we seek $x_0 \in X$ such that

$$f(x_0) = \inf_{x \in C} f(x)$$

where

$$C = \{x : g(x) = 0\}.$$

Provided that the functions are smooth enough and the constraints are *regular*, there is a very satisfactory Lagrange multiplier representation for this problem.

Definition 5.4.1. *Let X and Y be Banach spaces. Suppose $g : X \to Y$ is Fréchet differentiable on an open set $\mathcal{O} \subset X$, and suppose that the Fréchet derivative $Dg(x_0)$ is continuous at a point $x_0 \in \mathcal{O}$ in the uniform topology on the space $\mathcal{L}(X, Y)$. The point $x_0 \in \mathcal{O}$ is a regular point of the function f if $Dg(x_0)$ maps X onto Y.*

Ljusternik's Theorem

As will be seen in many applications, the regularity of the constraints plays an important role in justifying the applicability of Lagrange multipliers to many equality constrained problems. In fact, this pivotal role is made clear in the following theorem due to Ljusternik.

Theorem 5.4.1 (Ljusternik's Theorem). *Let X and Y be Banach spaces. Suppose that*

(i) *$g : X \to Y$ is Fréchet differentiable on an open set $\mathcal{O} \subseteq X$,*
(ii) *g is regular at $x_0 \in \mathcal{O}$, and*
(iii) *the Fréchet derivative $x_0 \mapsto Df(x_0)$ is continuous at x_0 in the uniform operator topology on $\mathcal{L}(X, Y)$.*

Then there is a neighborhood $\mathcal{N}(y_0)$ of $y_0 = g(x_0)$ and a constant C such that the equation

$$y = g(x)$$

has a solution x for every $y \in \mathcal{N}(y_0)$ and

$$\|x - x_0\|_X \leq C\|y - y_0\|_Y.$$

With these preliminary definitions, we can now state the Lagrange multiplier theorem for equality constrained extremization.

Theorem 5.4.2. *Let X and Y be Banach spaces, $f : X \to \mathbb{R}$, and $g : X \to Y$. Suppose that:*

(i) *f and g are Fréchet differentiable on an open set $\mathcal{O} \subseteq X$,*
(ii) *the Fréchet derivatives*

$$x_0 \mapsto Df(x_0)$$
$$x_0 \mapsto Dg(x_0)$$

are continuous in the uniform operator topology on $\mathcal{L}(X, \mathbb{R})$ and $\mathcal{L}(X, Y)$, respectively, and

(iii) *$x_0 \in \mathcal{O}$ is a regular point of the constraints $g(x)$.*

If f has a local extremum under the constraint $g(x_0) = 0$ at the regular point $x_0 \in \mathcal{O}$, then there is a Lagrange multiplier $y_0^* \in Y^*$ such that the Lagrangian

$$f(x) + y_0^* g(x)$$

is stationary at x_0. That is, we have

$$Df(x_0) + y_0^* \circ Dg(x_0) = 0.$$

Proof. We first show that if x_0 is a local extremum, then $Df(x_0) \circ x = 0$ for all x such that $Dg(x_0) \circ x = 0$. Define the mapping

$$F : X \to \mathbb{R} \times Y$$
$$F(x) = \big(f(x), g(x)\big).$$

Suppose, to the contrary, that there is $u \in X$ such that

$$Dg(x_0) \circ u = 0$$

but

$$Df(x_0) \circ u = z \neq 0.$$

If this were the case, then x_0 would be a regular point of the mapping F. To see why this is the case, we can compute

$$DF(x_0) \circ x = \left(Df(x_0) \circ x, Dg(x_0) \circ x\right) \in \mathbb{R} \times Y.$$

By assumption $Dg(x_0) : X \rightarrow Y$ is onto Y since x_0 is a regular point of the constraint $g(x) = 0$. Pick some arbitrary $(\alpha, y) \in \mathbb{R} \times Y$. Since $Dg(x_0)$ is onto Y, there is an $\bar{x} \in X$ such that

$$Dg(x_0) \circ \bar{x} = y.$$

The derivatives $Df(x_0) \circ u$ and $Dg(x_0) \circ u$ are linear in the increment u by definition. This fact, along with the definition of the real number β, where $\beta = Df(x_0) \circ \bar{x}$, implies that

$$Df(x_0) \circ \left(\frac{\alpha - \beta}{z} u\right) = \left(\frac{\alpha - \beta}{z}\right) Df(x_0) \circ u$$
$$= \alpha - \beta = \alpha - Df(x_0) \circ \bar{x}$$

$$Dg(x_0) \circ \left(\frac{\alpha - \beta}{z} u\right) = \left(\frac{\alpha - \beta}{z}\right) Dg(x_0) \circ u = 0.$$

If we choose $x = \left(\frac{\alpha - \beta}{z} u\right) + \bar{x}$, it is readily seen that

$$\begin{aligned}
DF(x_0) \circ \left(\frac{\alpha - \beta}{z} u + \bar{x}\right) &= \left(Df(x_0) \circ \left(\frac{\alpha - \beta}{z} u + \bar{x}\right), Dg(x_0) \circ \left(\frac{\alpha - \beta}{z} u + \bar{x}\right)\right) \\
&= \left(\frac{\alpha - \beta}{z} Df(x_0) \circ u + Df(x_0) \circ \bar{x}, Dg(x_0) \circ \bar{x}\right) \\
&= (\alpha - Df(x_0) \circ \bar{x} + Df(x_0) \circ \bar{x}, y) \\
&= (\alpha, y).
\end{aligned}$$

The map $DF(x_0)$ is consequently onto $\mathbb{R} \times Y$ and x_0 is a regular point of the map F. Define

$$F(x_0) = \left(f(x_0), g(x_0)\right) = (\alpha_0, 0) \in \mathbb{R} \times Y.$$

By Ljusternik's theorem, there is a neighborhood of $(\alpha_0, 0)$

$$\mathcal{N}(\alpha_0, 0) \subseteq \mathbb{R} \times Y$$

such that the equation

$$F(x) = (\alpha, y)$$

has a solution for every $(\alpha, y) \in \mathcal{N}(\alpha_0, 0)$ and the solution satisfies

$$\|x - x_0\|_X \leq C\left\{|\alpha - \alpha_0| + \|y\|_Y\right\}.$$

In particular, the element $(\alpha_0 - \epsilon, 0)$ is in the neighborhood $\mathcal{N}(\alpha_0, 0)$ for all ϵ small enough. For every $\epsilon > 0$ there is a solution x_ϵ to the equation

$$F(x_\epsilon) = (\alpha_0 - \epsilon, 0).$$

But this means that

$$f(x_\epsilon) = \alpha_0 - \epsilon$$
$$= f(x_0) - \epsilon$$

and

$$g(x_\epsilon) = 0.$$

Furthermore, we have that

$$\|x_\epsilon - x_0\|_X \leq \epsilon.$$

This contradicts the fact that x_0 is a local extremum, and we conclude

$$Df(x_0) \circ x = 0$$

for all $x \in X$ such that

$$Dg(x_0) \circ x = 0.$$

Recall that

$$\{x \in X : Dg(x_0) \circ x = 0\} = \ker\big(Dg(x_0)\big).$$

In fact $Df(x_0) \in X^*$ and $Df(x_0) \in \big(\ker\left(Dg(x_0)\right)\big)^\perp$. Since the range of $Dg(x_0)$ is closed, we have

$$\mathrm{range}\big((Dg(x_0))^*\big) = \big(\ker(Dg(x_0))\big)^\perp.$$

By definition

$$Dg(x_0) : X \to Y$$

and

$$(Dg(x_0))^* : Y^* \to X^*.$$

We conclude that there is a $y_0^* \in Y^*$ such that

$$Df(x_0) = -(Dg(x_0))^* \circ y_0^*$$
$$Df(x_0) + (Dg(x_0))^* \circ y_0^* = 0.$$

By definition

$$\big\langle (Dg(x_0))^* \circ y_0^*, x \big\rangle_{X^* \times X} = \langle y_0^*, Dg(x_0) \circ x \rangle_{Y^* \times Y}.$$

So that this equality can be written as

$$Df(x_0) + y_0^* \circ Dg(x_0) = 0. \qquad \square$$

The above theorem bears a close resemblance to the Lagrange multiplier theorem from undergraduate calculus discussed in [12], [18] in the introduction. The essential ingredients of the above theorem include smoothness of the functionals f and g and the regularity of the constraints. There is an alternative form of this theorem that weakens the requirement that the constraints are in fact regular at x_0. It will be useful in many applications.

Theorem 5.4.3. *Let X and Y be Banach spaces, $f : X \to \mathbb{R}$, and $g : X \to Y$. Suppose that:*

(i) *f and g are Fréchet differentiable on an open set $\mathcal{O} \subseteq X$,*

(ii) *the Fréchet derivatives*

$$x_0 \mapsto Df(x_0)$$
$$x_0 \mapsto Dg(x_0)$$

are continuous in the uniform operator topology on $\mathcal{L}(X, \mathbb{R})$ and $\mathcal{L}(X, Y)$, respectively, and

(iii) *the range of $Dg(x_0)$ is closed in Y.*

If f has a local extremum under the constraint $g(x_0) = 0$ at the point $x_0 \in \mathcal{O}$, then there are multipliers $\lambda_0 \in \mathbb{R}$ and $y_0^* \in Y^*$ such that the Lagrangian

$$\lambda_0 f(x) + y_0^* g(x)$$

is stationary at x_0. That is

$$\lambda_0 Df(x_0) + y_0^* \circ Dg(x_0) = 0.$$

Proof. The proof of this theorem can be carried out in two steps. First, suppose that the range of $Dg(x_0)$ is all of Y. In this case, the constraint $g(x_0)$ is regular at x_0. We can apply the preceding theorem and select $\lambda_0 \equiv 1$. If, on the other hand, the range $Dg(x_0)$ is strictly contained in Y, we know that there is some $\tilde{y} \in Y$ such that

$$d = \inf \left\{ \|\tilde{y} - y\| : y \in \text{range}(Dg(x_0)) \right\} > 0.$$

By Theorem 2.2.2 there is an element $y_0^* \in \left(\text{range}(Dg(x_0)) \right)^{\perp}$ such that

$$\langle y_0^*, \tilde{y} \rangle = d \neq 0$$

and $y_0^* \neq 0$. But for any linear operator A

$$\left(\text{range}(A) \right)^{\perp} = \ker(A^*)$$

so that

$$y_0^* \in \left(\text{range}(Dg(x_0)) \right)^{\perp} \equiv \ker \left((Dg(x_0))^* \right).$$

By definition, since $y_0^* \in Y^*$

$$\langle y_0^*, Dg(x_0) \circ x \rangle_{Y^* \times Y} = \left\langle (Dg(x_0))^* \circ y_0^*, x \right\rangle_{X^* \times X} = 0$$

for all $x \in X$. We choose $\lambda_0 = 0$ and conclude

$$\lambda_0 Df(x_0) + y_0^* \circ Dg(x_0) = 0. \qquad \square$$

5.5   Fréchet Differentiable Implicit Functionals

In all of our discussions so far in this chapter, optimization problems having quite general forms have been considered. In this section, we discuss a class of optimization problem that has a very specific structure. These problems often arise in the study of optimal control. It has been noted by several authors that the distinguishing feature of optimal control problems within the field of optimization is their distinct structure. The standard optimization problem is such that we seek $u_0 \in \mathcal{U}$ such that

$$J(u_0) = \inf_u \{J(u) : u \in \mathcal{U}\}. \tag{5.4}$$

Instead of Equation (5.4), optimal control problems frequently arise where we seek a pair $(x_0, u_0) \in X \times \mathcal{U}$ such that

$$\mathcal{J}(x_0, u_0) = \inf_{x, u} \{\mathcal{J}(x, u) : A(x, u) = 0, u \in \mathcal{U}\}. \tag{5.5}$$

In these equations, x represents the dependent quantity or physical state of the system under consideration, while u denotes the input or control. Optimization problems having the form depicted in Equation (5.5) arise in control problems for physical reasons. We typically seek to minimize some quantity such as fuel, cost, departure motion, or vibration, etc. subject to a collection governing equations that are inviolate. In Equation (5.5),

$$A(x, u) = 0 \tag{5.6}$$

denotes the equations of physics that relate the inputs u to the states x. It is a fundamental premise of optimal control that a pair $(x, u) \in X \times \mathcal{U}$ that does not satisfy Equation (5.6) violates some physical law. The equations of evolution that govern the physical variables of the problem are encoded by ordinary differential equations, partial differential equations or integral equations represented by Equation (5.6). It is usually very difficult, either computationally or analytically to solve for the state x as a function of the control u in Equation (5.6). If we can find $x(u)$ that satisfies Equation (5.6)

$$A(x(u), u) = 0$$

it is clear that we can reduce Equation (5.5) to the form in Equation (5.4).

$$J(u) \triangleq \mathcal{J}(x(u), u)$$

$$J(u_0) = \mathcal{J}(x(u_0), u_0) = \inf_u \{J(u) = \mathcal{J}(x(u), u) : u \in \mathcal{U}\}.$$

In some cases, it will be possible to solve for $x(u)$. More frequently, it will not. We will need methods for calculating the Gateaux derivative of $J(u)$ without calculating $x(u)$ explicitly. This task is accomplished by using the co-state, adjoint, or optimality system equations.

The following theorem provides the theoretical foundation of adjoint, co-state or optimality system methods for Fréchet differentiable, implicit functionals.

Theorem 5.5.1. *Let X, Y, U be normed vector spaces, $\mathcal{J} : X \times U \to Z$, and suppose $A : X \times U \to Y$ defines a unique function $x(u)$ via the solution of*

$$A(x(u), u) = 0.$$

Suppose further that:

- *$A(\cdot, u) : X \to Y$ is Fréchet differentiable at $x = x(u)$,*
- *$\mathcal{J}(\cdot, u) : X \to Z$ is Fréchet differentiable at $x = x(u)$,*
- *$A(x, \cdot) : U \to Y$ is Gateaux differentiable,*
- *$\mathcal{J}(x, \cdot) : U \to Z$ is Gateaux differentiable,*
- *Gateaux differential $D_u A(x, u)$ is continuous on $X \times U$,*
- *Gateaux differential $D_u \mathcal{J}(x, u)$ is continuous on $X \times U$, and*
- *$x(u)$ is Lipschitz continuous*

$$\|x(u) - x(v)\|_X \leq C\|u - v\|_U$$

   for some $C \in \mathbb{R}$.

If there is a solution $\lambda \in \mathcal{L}(Y, Z)$ to the equation

$$\lambda \circ D_x A(x, u) = D_x \mathcal{J}(x, u)$$

at $x = x(u)$, then

$$J(u) \overset{\triangle}{=} \mathcal{J}(x(u), u)$$

is Gateaux differentiable at u and

$$DJ(u) = D_u \mathcal{J}(x, u) - \lambda \circ D_u A(x, u) \in \mathcal{L}(U, Z).$$

Proof. Suppose $0 \leq \epsilon \leq 1$. For $u, \tilde{u} \in U$, define

$$
\begin{aligned}
u_\epsilon &\overset{\triangle}{=} u + \epsilon(\tilde{u} - u) \quad \in U \\
x_\epsilon &\overset{\triangle}{=} x(u_\epsilon).
\end{aligned}
\tag{5.7}
$$

Recall that we want to find an expression for $DJ(u)$

$$DJ(u) \circ v \overset{\triangle}{=} \lim_{\epsilon \to 0} \frac{J(u + \epsilon v) - J(u)}{\epsilon}. \tag{5.8}$$

We have

$$
\begin{aligned}
\frac{J(u_\epsilon) - J(u)}{\epsilon} &= \frac{\mathcal{J}(x(u_\epsilon), u_\epsilon) - \mathcal{J}(x(u), u)}{\epsilon} \\
&= \frac{\mathcal{J}(x(u_\epsilon), u_\epsilon) - \mathcal{J}(x(u_\epsilon), u)}{\epsilon} + \frac{\mathcal{J}(x(u_\epsilon), u) - \mathcal{J}(x(u), u)}{\epsilon}.
\end{aligned}
\tag{5.9}
$$

By the Fréchet differentiability of $\mathcal{J}(\cdot, u)$

$$\mathcal{J}(x(u_\epsilon), u) = \mathcal{J}(x(u), u) + D_x \mathcal{J}(x(u), u) \circ (x(u_\epsilon) - x(u)) + R_1\big(\|x(u_\epsilon) - x(u)\|_X\big)$$

where the remainder $R_1(\cdot)$ satisfies

$$\frac{R_1\left(\|x(u_\epsilon) - x(u)\|_X\right)}{\|x(u_\epsilon) - x(u)\|_X} \to 0$$

as

$$\|x(u_\epsilon) - x(u)\|_X \to 0.$$

By the Lipschitz continuity of $x(u)$, we note that

$$\|x(u_\epsilon) - x(u)\| \le C\|\tilde{u} - u\|_U \cdot \epsilon$$

so that

$$\frac{R_1\left(\|x(u_\epsilon) - x(u)\|_X\right)}{C\|\tilde{u} - u\|_U \cdot \epsilon} \le \frac{R_1\left(\|x(u_\epsilon) - x(u)\|_X\right)}{\|x(u_\epsilon) - x(u)\|_X}.$$

Consequently, we write

$$\mathcal{J}(x(u_\epsilon), u) = \mathcal{J}(x(u), u) + D_x\mathcal{J}(x(u), u) \circ (x(u_\epsilon) - x(u)) + R(\epsilon) \qquad (5.10)$$

where $R(\epsilon)$ is a remainder term such that

$$\lim_{\epsilon \to 0} \frac{R(\epsilon)}{\epsilon} \to 0.$$

In the various derivative expressions that follow, we will use $R(\epsilon)$ to denote generically any remainder terms that have the above asymptotic behavior as a function of ϵ. In addition, by the Gateaux differentiability of $\mathcal{J}(x, \cdot)$, we have

$$\begin{aligned} \mathcal{J}(x(u_\epsilon), u_\epsilon) &= \mathcal{J}(x(u_\epsilon), u) + D_u\mathcal{J}(x(u_\epsilon), u) \circ (u_\epsilon - u) + R(\epsilon) \\ &= \mathcal{J}(x(u_\epsilon), u) + \epsilon D_u\mathcal{J}(x(u_\epsilon), u) \circ (\tilde{u} - u) + R(\epsilon). \end{aligned} \qquad (5.11)$$

Substituting Equations (5.10) and (5.11) into (5.9) yields

$$\begin{aligned} \frac{J(u_\epsilon) - J(u)}{\epsilon} &= D_u\mathcal{J}(x(u_\epsilon), u) \circ (\tilde{u} - u) \\ &+ \frac{D_x\mathcal{J}(x(u), u) \circ (x(u_\epsilon) - x(u))}{\epsilon} + \frac{R(\epsilon)}{\epsilon}. \end{aligned} \qquad (5.12)$$

Since the pairs $(x(u_\epsilon), u_\epsilon)$, $(x(u), u)$ are solutions of $A(\cdot, \cdot) = 0$, it is always true that

$$\lambda \circ (A(x(u_\epsilon), u_\epsilon) - A(x(u), u)) = 0 \qquad \in Z.$$

We can write

$$\lambda \circ \left\{ \frac{A(x(u_\epsilon), u_\epsilon) - A(x(u_\epsilon), u)}{\epsilon} + \frac{A(x(u_\epsilon), u) - A(x(u), u)}{\epsilon} \right\} = 0. \qquad (5.13)$$

Since A is Gateaux differentiable in its second argument

$$A(x(u_\epsilon), u_\epsilon) = A(x(u_\epsilon), u) + D_u A(x(u_\epsilon), u) \circ (u_\epsilon - u) + R(\epsilon)$$
$$= A(x(u_\epsilon), u) + \epsilon D_u A(x(u_\epsilon), u) \circ (\tilde{u} - u) + R(\epsilon) \tag{5.14}$$

and Fréchet differentiable in its first argument

$$A(x(u_\epsilon), u) = A(x(u), u) + D_x A(x(u), u) \circ (x(u_\epsilon) - x(u)) + R_2(\|x(u_\epsilon) - x(u)\|)$$
$$= A(x(u), u) + D_x A(x(u), u) \circ (x(u_\epsilon) - x(u)) + R(\epsilon). \tag{5.15}$$

When we substitute Equations (5.15) and (5.14) into (5.13), we obtain

$$\lambda \circ \left\{ D_u A(x(u_\epsilon), u) \circ (\tilde{u} - u) + \frac{D_x A(x(u), u) \circ (x(u_\epsilon) - x(u))}{\epsilon} + \frac{R(\epsilon)}{\epsilon} \right\} = 0. \tag{5.16}$$

By hypothesis, we have

$$\lambda \circ D_x A(x(u), u) \circ (x(u_\epsilon) - x(u)) = D_x J(x(u), u) \circ (x(u_\epsilon) - x(u))$$

which, from Equation (5.16) implies that

$$\frac{D_x J(x(u), u) \circ (x(u_\epsilon) - x(u))}{\epsilon} = -\lambda \circ \left\{ D_u A(x(u_\epsilon), u) \circ (\tilde{u} - u) + \frac{R(\epsilon)}{\epsilon} \right\}.$$

When we substitute this expression into Equation (5.12), we obtain

$$\frac{J(u_\epsilon) - J(u)}{\epsilon} = D_u J(x(u_\epsilon), u) \circ (\tilde{u} - u) - \lambda \circ D_u A(x(u_\epsilon), u) \circ (\tilde{u} - u) + \frac{R(\epsilon)}{\epsilon}.$$

Recalling that $u_\epsilon = u + \epsilon(\tilde{u} - u)$, we can take the limit above to obtain

$$DJ(u) \circ (\tilde{u} - u) = D_u J(x(u), u) \circ (\tilde{u} - u) - \lambda \circ D_u A(x(u), u) \circ (\tilde{u} - u).$$

In this last limit, we have used the continuity of $D_u J(x, u)$ and $D_u A(x, u)$ on $X \times U$. $\qquad \square$

Chapter 6

Convex Functionals

Mathematicians worked diligently to extend the "prototypical" optimization results of the calculus of a real variable to a broader functional analytic setting. The result was the emergence of the field of convex analysis [7], [25]. In this section, we review the fundamentals of convex analysis, and study its relationship to continuity and differentiability properties of functionals defined on normed vector spaces.

As usual, the definition of a convex set agrees with our intuitive understanding of what constitutes convexity. Simply put, a set can be thought to be convex if it has no reentrant "bumps or corners." Mathematically, this idea can be made precise by defining the *segment* joining any two points x and y in a vector space to be the set of all points having the form $\alpha x + (1 - \alpha)y$ for $0 \leq \alpha \leq 1$. A set is convex if and only if it contains all the segments joining any two of its points.

Definition 6.0.1. *A set C is convex if and only if*

$$x, y \in C \quad \Longrightarrow \quad \alpha x + (1 - \alpha)y \in C \quad \forall \alpha \in [0, 1].$$

The geometric nature of this definition is depicted in Figure 6.1. Given a collection of convex sets, it is always possible to create other convex sets by the operations of *dilation, sum* or *intersection*.

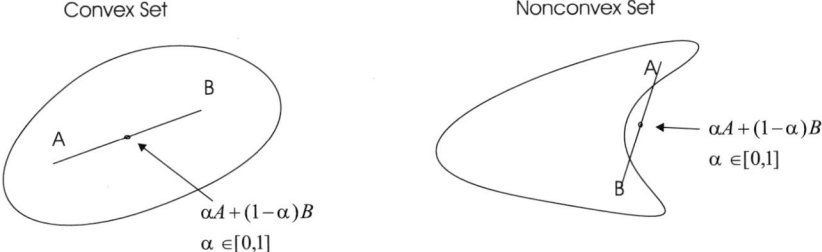

Figure 6.1: Convex and Nonconvex Sets

Theorem 6.0.2. *If A and B are convex sets then*

1. *The dilation λA of a convex set A is convex. In other words, the set*

$$\lambda A = \{\lambda x : x \in A\}$$

is convex.

2. *The sum $A + B$ of convex sets is convex. In other words, the set*

$$\{a + b : a \in A, b \in B\}$$

is convex.

3. *The arbitrary intersection of convex sets is convex.*

Proof. The proof that these three operations yield convex sets is quite simple. The dilation of a convex set is convex: Let λ be fixed in $\mathbb{R}$. Pick any pair of points $x, y \in \lambda A$ where A is convex. Then there exists $x_A \in A$ such that $\lambda x_A = x$, and $y_A \in A$ such that $\lambda y_A = y$. Since A is convex, we have that

$$\alpha x_A + (1 - \alpha) y_A = z_A \in A.$$

Multiplying this equation by λ, it follows that

$$\lambda \alpha x_A \;\; + \;\; \lambda(1 - \alpha) y_A = \lambda z_A$$
$$\alpha x \;\; + \;\; (1 - \alpha) y = z \in \lambda A \quad \forall \alpha \in [0, 1].$$

The sum of convex sets is convex: Let $x + y \in A + B$, where A, B are convex. Then there exists points x_A, x_B, y_A, y_B such that

$$x \;\; = \;\; x_A + x_B$$
$$y \;\; = \;\; y_A + y_B$$

where $x_A, y_A \in A$ and $x_B, y_B \in B$. Since A and B are convex we have

$$\alpha x_A \;\; + \;\; (1 - \alpha) y_A \in A$$
$$\alpha x_B \;\; + \;\; (1 - \alpha) y_B \in B \quad \forall \alpha \in [0, 1].$$

Adding these equations, we obtain

$$\alpha x + (1 - \alpha) y = (\alpha x_A + (1 - \alpha) y_A) + (\alpha x_B + (1 - \alpha) y_B) \quad \in A + B$$

for all $\alpha \in [0, 1]$.

The arbitrary intersection of convex sets is convex: Suppose that $\{A_k\}$ is a family of convex sets, and select $x, y \in \bigcap_k A_k$. By definition, for each k we have that $\alpha x + (1 - \alpha) y \in A_k$ for each $\alpha \in [0, 1]$. It follows that $\alpha x + (1 - \alpha) y \in \bigcap_k A_k$ for each $\alpha \in [0, 1]$, and that the set $\bigcap_k A_k$ is convex. $\square$

As we finish this section introducing convex sets, we note three special classes of convex sets that will play a special role in the remainder of this chapter, *hyperplanes*, *cones* and *convex hulls*.

Definition 6.0.2. *A linear variety, or hyperplane, V is a translation of a subspace U of linear vector space X. In other words, if V is a linear variety in a vector space X, then every element $v \in V$ can be expressed as*

$$v = v_0 + u \quad \text{for some } u \in U$$

where $v_0 \in X$ is the fixed translation, and U is a subspace of the vector space X.

Definition 6.0.3. *Let A be an arbitrary set in a vector space X. The convex hull of A, denoted $\mathrm{co}(A)$, is the smallest convex set containing A. In other words, the convex hull of an arbitrary set A is the intersection of all convex sets containing A.*

6.1 Characterization of Convexity

As noted in the introduction of this chapter, convexity becomes a particularly useful property for studying extremal problems when

- the set of constraints is convex, or
- the graph of the function to be minimized is convex.

In the last section, some rudiments for ascertaining if a given set is convex have been introduced. In this section, we show that functions defined on vector spaces can likewise be given a useful notion of convexity. By convention, convex analysis is often couched in terms of functionals defined on the extended real line $\overline{\mathbb{R}} \equiv \mathbb{R} \cup \{\infty\}$.

Definition 6.1.1. *An extended functional f on a vector space X is a mapping*

$$f : X \to \overline{\mathbb{R}}.$$

The extended functional is said to be strict, or proper if it is not identically equal to $+\infty$.

As usual, we identify a functional $f : X \to \overline{\mathbb{R}}$ with its graph

$$\mathrm{graph}(f) = \{(x, y) \in X \times \mathbb{R} : y = f(x)\}$$

as shown in Figure 6.2 for $X \subseteq \mathbb{R}$. The (closed) lower sections of the graph of f are defined to be the inverse images under f of any set of the form $(-\infty, r]$ for some $r \in \mathbb{R}$. An example of the lower sections of the graph of f for a fixed $r \in \mathbb{R}$ are depicted schematically in Figure 6.3.

Definition 6.1.2. *A functional is convex if and only if*

$$f(\alpha x + (1 - \alpha)y) \leq \alpha f(x) + (1 - \alpha)f(y) \quad \forall \alpha \in [0, 1].$$

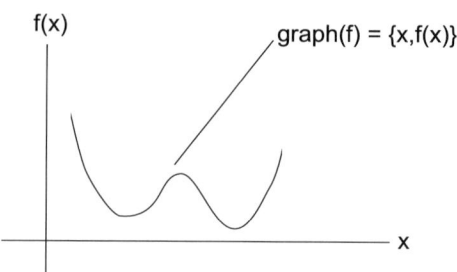

Figure 6.2: Graph of a Function

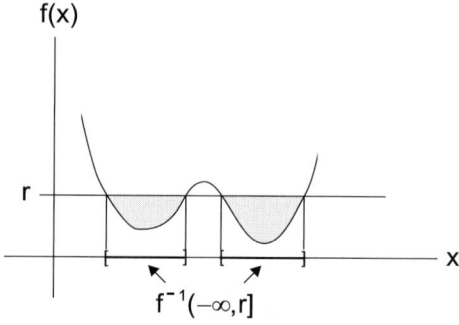

Figure 6.3: Lower Sections of the Graph

Thus, a functional is convex if the segment connecting two points on its graph lies above the graph of the function, as shown in Figure 6.4. Now, this definition is often expressed concisely by defining the set of all points that lie above the graph of a extended functional to be the *epigraph*.

Definition 6.1.3. *The epigraph of a functional f acting on a vector space X is the subset*

$$\mathrm{epi}(f) = \{(x, \gamma) \in X \times \mathbb{R} : f(x) \leq \gamma\}.$$

The following theorem makes clear how the definitions of convexity of a set and convexity of a functional are related.

Proposition 6.1.1. *Suppose f is a functional defined on a convex set. Then f is a convex functional if and only if the set $\mathrm{epi}(f)$ is convex.*

Proof. First, suppose that f is convex, and consider two points in the epigraph of f,

$$
\begin{aligned}
(x, \xi) &\in \mathrm{epi}(f) \subseteq X \times \mathbb{R} \\
(y, \eta) &\in \mathrm{epi}(f) \subseteq X \times \mathbb{R}.
\end{aligned}
$$

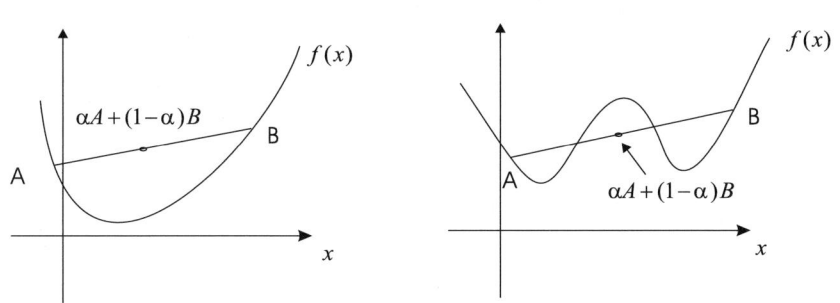

Figure 6.4: Convex and Nonconvex Functionals

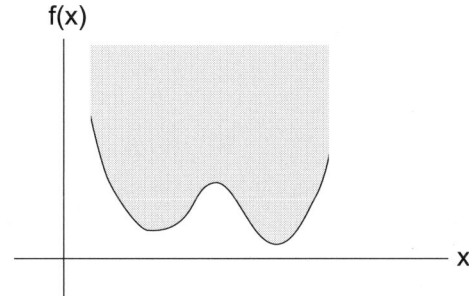

Figure 6.5: Epigraph of a Functional

To prove sufficiency, it is enough to show that

$$\alpha(x, \xi) + (1 - \alpha)(y, \eta) \in \text{epi}(f) \quad \forall \alpha \in [0, 1].$$

But by the convexity of f, we can write

$$f(\alpha x + (1 - \alpha)y) \leq \alpha f(x) + (1 - \alpha)f(y) \quad \forall \alpha \in [0, 1] \tag{6.1}$$
$$\leq \alpha \xi + (1 - \alpha)\eta \tag{6.2}$$

where the latter inequality follows from the fact that $(x, \xi) \in \text{epi}(f)$ and $(y, \eta) \in \text{epi}(f)$. But Equations (6.1)–(6.2), simply say that

$$(\alpha x + (1 - \alpha)y, \alpha \xi + (1 - \alpha)\eta) \in \text{epi}(f)$$

or

$$\alpha(x, \xi) + (1 - \alpha)(y, \eta) \in \text{epi}(f).$$

To prove necessity, suppose that $\text{epi}(f)$ is convex. By definition, we have that

$$(x, f(x)) \in \text{epi}(f)$$
$$(y, f(y)) \in \text{epi}(f).$$

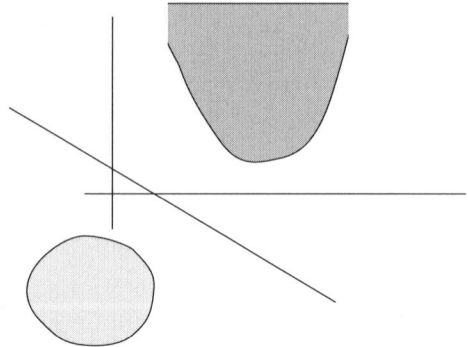

Figure 6.6: Separation of Sets via a Hyperplane

By the convexity of epi(f), we can write

$$\alpha(x, \xi) + (1 - \alpha)(y, \eta) \in \text{epi}(f) \quad \forall \alpha \in [0, 1] \tag{6.3}$$

for any $(x, \xi), (y, \eta) \in \text{epi}(f)$. In particular, for $(x, \xi) = (x, f(x))$ and $(y, \eta) = (y, f(y))$ Equation (6.3) becomes

$$\alpha(x, f(x)) + (1 - \alpha)(y, f(y)) \in \text{epi}(f)$$
$$(\alpha x + (1 - \alpha)y, \alpha f(x) + (1 - \alpha)f(y)) \in \text{epi}(f).$$

But this equation just asserts that

$$f(\alpha x + (1 - \alpha)y) \leq \alpha f(x) + (1 - \alpha)f(y) \quad \forall \alpha \in [0, 1]$$

and f is convex.   □

6.2   Gateaux Differentiable Convex Functionals

From a practical standpoint then, our alternatives for establishing the existence of solutions to minimization problems are enhanced if we can establish that a functional is convex. The following theorem provides several equivalent characterizations of convexity, if a functional is Gateaux differentiable. The utility of this theorem rests in the fact that the conditions required to prove convexity arise frequently in applications.

Theorem 6.2.1. *Let X be a reflexive Banach space, and let $f : M \subseteq X \to \overline{\mathbb{R}}$ be Gateaux differentiable over the closed, convex set M. Then the following conditions are equivalent:*

(i) *f is convex over M.*

(ii) *We have*

$$f(u) - f(v) \geq \langle Df(v), u - v \rangle_{X^* \times X}, \quad \forall u, v \in M.$$

(iii) *The first Gateaux derivative is monotone,*

$$\langle Df(u) - Df(v), u - v \rangle_{X^* \times X} \geq 0, \quad \forall\, u, v \in M.$$

(iv) *The second Gateaux derivative of f exists and is positive,*

$$\langle D^2 f(u) \circ v, v \rangle_{X^* \times X} \geq 0 \quad \forall\, v \in M.$$

Proof. (i) $\Rightarrow$ (ii). Because f is convex, we have

$$
\begin{aligned}
\lambda f(u) + (1 - \lambda) f(v) &\geq f(\lambda u + (1 - \lambda) v) \quad \forall\, u, v \in M \\
\lambda (f(u) - f(v)) &\geq f(v + \lambda(u - v)) - f(v) \quad \forall\, u, v \in M. \qquad (6.4)
\end{aligned}
$$

Dividing Equation (6.4) by λ, we obtain

$$
\begin{aligned}
f(u) - f(v) &\geq \frac{f(v + \lambda(u - v)) - f(v)}{\lambda} \\
&\geq \langle Df(v), u - v \rangle_{X^* \times X}. \qquad (6.5)
\end{aligned}
$$

(ii) $\Rightarrow$ (iii). By interchanging the roles of u and v in Equation (6.5), we can also write

$$f(v) - f(u) \geq \langle Df(u), v - u \rangle_{X^* \times X} = -\langle Df(u), u - v \rangle_{X^* \times X}. \qquad (6.6)$$

By adding equations (6.5) and (6.6), we obtain

$$\langle Df(u) - Df(v), u - v \rangle_{X^* \times X} \geq 0.$$

(iii) $\Rightarrow$ (iv). Now let us suppose that the first Gateaux derivative of f is monotone, and show that this implies that the second Gateaux derivative is positive. By definition, we know that

$$
\begin{aligned}
Df(u, v) &= \lim_{\epsilon \to 0} \frac{f(u + \epsilon v) - f(u)}{\epsilon} \\
D^2 f(u, v, w) &= \lim_{\epsilon \to 0} \frac{Df(u + \epsilon w, v) - Df(u, v)}{\epsilon}. \qquad (6.7)
\end{aligned}
$$

Let us consider just the numerator of the right-hand side in equation (6.7). Suppose we choose, $w = v$. Then we can write

$$
\begin{aligned}
\epsilon \big(Df(u + \epsilon v) \circ v - Df(u) \circ v \big) &= \big(Df(u + \epsilon v) - Df(u) \big) \circ (\epsilon v) \\
&= \big(Df(u + \epsilon v) - Df(u) \big) \circ \big((u + \epsilon v) - u \big) \\
&= \langle Df(u + \epsilon v) - Df(u), (u + \epsilon v) - u \rangle_{X^* \times X} \\
&\geq 0.
\end{aligned}
$$

This last inequality follows from the monotonicity in (iii) above. Hence, by the monotonicity of Df, we know that

$$
\begin{aligned}
\langle D^2 f(u) \circ v, v \rangle_{X^* \times X} &= D^2 f(u, v, v) \\
&= \lim_{\epsilon \to 0} \frac{Df(u + \epsilon v, v) - Df(u, v)}{\epsilon} \\
&\geq 0.
\end{aligned}
$$

To complete the proof, we must show that

(iv) $\Rightarrow$ (iii). Suppose that

$$\langle D^2 f(u) \circ v, v \rangle_{X^* \times X} \geq 0 \quad \forall u, v \in M.$$

Let $u = u(t) = u_1 + t(u_2 - u_1)$ and $v = (u_2 - u_1)$ for $u_1, u_2 \in M$ and $t \in [0, 1]$. Introducing a function $\phi(t) = Df(u(t), u_2 - u_1)$, we rewrite the last inequality as

$$\langle D^2 f(u(t)) \circ (u_2 - u_1), (u_2 - u_1) \rangle_{X^* \times X} = \lim_{dt \to 0} \frac{\phi(t + dt) - \phi(t)}{dt} \geq 0 \quad \forall t \in [0, 1]$$

which is reduced to

$$\phi(t + dt) \geq \phi(t) \quad \forall t \in [0, 1].$$

It follows that the function $\phi(t)$ is nondecreasing on $[0, 1]$. Consequently, $\phi(1) \geq \phi(0)$, or, equivalently,

$$\phi(1) = \langle Df(u_2), u_2 - u_1 \rangle \geq \langle Df(u_1), u_2 - u_1 \rangle = \phi(0).$$

(iii) $\Rightarrow$ (ii). Now suppose that

$$\langle Df(u) - Df(v), u - v \rangle_{X^* \times X} \geq 0 \quad \forall u, v \in M.$$

For $u = u(t) = u_1 + t(u_2 - u_1)$ and $v = u_1$, where $u_1, u_2 \in M$ and $t \in [0, 1]$, this inequality is rewritten as

$$\langle Df(u(t)), u_2 - u_1 \rangle_{X^* \times X} \geq \langle Df(u_1), u_2 - u_1 \rangle_{X^* \times X} \quad \forall t \in [0, 1]$$

which, by definition, is equivalent to

$$\lim_{dt \to 0} \frac{f(u(t + dt)) - f(u(t))}{dt} \geq \langle Df(u_1), u_2 - u_1 \rangle_{X^* \times X} \quad \forall t \in [0, 1]$$

or

$$f(u(t + dt)) - \langle Df(u_1), u_2 - u_1 \rangle_{X^* \times X} dt \geq f(u(t)) \quad \forall t \in [0, 1].$$

As a consequence, we obtain

$$f(u(1)) - \langle Df(u_1), u_2 - u_1 \rangle_{X^* \times X} \int_0^1 dt \geq f(u(0)).$$

But this means that

$$f(u_2) - f(u_1) \geq \langle Df(u_1), u_2 - u_1 \rangle_{X^* \times X} \quad \forall u_1, u_2 \in M.$$

(ii) $\Rightarrow$ (i). Finally, suppose that

$$f(u) - f(v) \geq \langle Df(v), u - v \rangle_{X^* \times X} \quad \forall u, v \in M.$$

For $u = u_1$ and $v = \lambda u_1 + (1 - \lambda)u_2$, $\lambda \in [0, 1]$, this inequality is reduced to

$$\frac{f(u_1) - f(\lambda u_1 + (1 - \lambda)u_2)}{1 - \lambda} \geq -\langle Df(\lambda u_1 + (1 - \lambda)u_2), u_2 - u_1\rangle_{X^* \times X} \quad (6.8)$$

and for $u = u_2$ and $v = \lambda u_1 + (1 - \lambda)u_2$, $\lambda \in [0, 1]$, we have

$$\frac{f(u_2) - f(\lambda u_1 + (1 - \lambda)u_2)}{\lambda} \geq \langle Df(\lambda u_1 + (1 - \lambda)u_2), u_2 - u_1\rangle_{X^* \times X}. \quad (6.9)$$

From (6.8) and (6.9), we obtain

$$\frac{f(u_2) - f(\lambda u_1 + (1 - \lambda)u_2)}{\lambda} \geq \frac{f(\lambda u_1 + (1 - \lambda)u_2) - f(u_1)}{1 - \lambda}$$

or, equivalently,

$$\lambda f(u_1) + (1 - \lambda)f(u_2) \geq f(\lambda u_1 + (1 - \lambda)u_2). \qquad \square$$

6.3 Convex Programming in $\mathbb{R}^n$

We will discuss multiplier methods for convex optimization problems that are progressively more general in this, and the following, sections. To begin, we limit consideration to optimization problems wherein we seek $x_0 \in \mathbb{R}^n$ such that

$$f(x_0) = \mu = \inf \{f(x) : g_i(x) \leq 0, \ x \in \mathbb{R}^n, \ i = 1, \ldots, m\}$$

and

$$g_i(x_0) \leq 0 \qquad \forall\, i = 1, \ldots, m$$

where

$$f : \quad \mathbb{R}^n \to \mathbb{R}$$
$$g_i : \quad \mathbb{R}^n \to \mathbb{R}.$$

This study will lead in subsequent sections to various generalizations for functionals defined on convex subsets of normed vector spaces.

Theorem 6.3.1. *Let* $f : \mathcal{U} \subseteq \mathbb{R}^n \to \mathbb{R}$ *and* $g_i : \mathcal{U} \subseteq \mathbb{R}^n \to \mathbb{R}$, $i = 1, \ldots, m$ *be proper convex functions where* $\mathcal{U}$ *is a convex set. Then there are nonnegative real numbers* α, β_i, $i = 1, \ldots, m$

$$\alpha \geq 0$$
$$\beta_i \geq 0$$

not all equal to zero such that

$$\alpha(f(x) - \mu) + \sum_{i=1}^{m} \beta_i g_i(x) \geq 0$$

for all $x \in \mathcal{U}$.

Proof. The proof of this theorem is based on an application of the Hahn-Banach theorem. Define the set

$$C \triangleq \{(f(x) - \mu + \gamma_0, g_1(x) + \gamma_1, \ldots, g_m(x) + \gamma_m) :$$
$$x \in \mathcal{U}, \ \gamma_i > 0, \ i = 1, \ldots, m\}$$
$$\subseteq \mathbb{R}^{m+1}.$$

We first note that the set C is convex. Indeed, if we let

$$c \triangleq (c_0, c_1, \ldots, c_m) \in C \subseteq \mathbb{R}^{m+1}$$
$$d \triangleq (d_0, d_1, \ldots, d_m) \in C \subseteq \mathbb{R}^{m+1}$$

then it is clear that

$$\begin{aligned}
\lambda c_0 + (1 - \lambda)d_0 &= \lambda(f(x_c) - \mu + \gamma_{c,0}) + (1 - \lambda)(f(x_d) - \mu + \gamma_{d,0}) \\
&= \lambda f(x_c) + (1 - \lambda)f(x_d) - \mu + \lambda\gamma_{c,0} + (1 - \lambda)\gamma_{d,0} \\
&\geq f(\lambda x_c + (1 - \lambda)x_d) - \mu + \lambda\gamma_{c,0} + (1 - \lambda)\gamma_{d,0}
\end{aligned}$$

by the convexity of f. In other words,

$$\lambda c_0 + (1 - \lambda)d_0 = f(\lambda x_c + (1 - \lambda)x_d) - \mu + \overline{\gamma_0}$$

for some $\overline{\gamma_0} > 0$. Likewise, for $i = 1, \ldots, m$,

$$\begin{aligned}
\lambda c_i + (1 - \lambda)d_i &= \lambda(g_i(x_c) + \gamma_{c,i}) + (1 - \lambda)(g_i(x_d) + \gamma_{d,i}) \\
&= \lambda g_i(x_c) + (1 - \lambda)g_i(x_d) + \lambda\gamma_{c,i} + (1 - \lambda)\gamma_{d,i} \\
&\geq g_i(\lambda x_c + (1 - \lambda)x_d) + \lambda\gamma_{c,i} + (1 - \lambda)\gamma_{d,i}
\end{aligned}$$

by the convexity of g_i, $i = 1, \ldots, m$. That is,

$$\lambda c_i + (1 - \lambda)d_i = g_i(\lambda x_c + (1 - \lambda)x_d) + \overline{\gamma_i}$$

for some $\overline{\gamma_i} > 0$. Thus, $c, d \in C$ implies that $\lambda c + (1 - \lambda)d \in C$ for all $0 \leq \lambda \leq 1$. Upon consideration, it is also evident that $0 \notin C$. If indeed it were the case that the origin is an element of C, we would have some $\hat{x}$ and $\gamma_0 > 0$ such that

$$f(\hat{x}) - \mu + \gamma_0 = 0$$

and

$$g_1(\hat{x}) \leq 0, \ g_2(\hat{x}) \leq 0, \ \ldots, \ g_m(\hat{x}) \leq 0$$

which contradicts the definition of μ as the infimum of f over the set of constrained elements. Consequently, the sets C and $\{0\}$ are a pair of nonempty, disjoint, convex sets. By the Hahn-Banach theorem, there are real numbers

$$\alpha \in \mathbb{R}, \quad \beta_i \in \mathbb{R}, \quad i = 1, \ldots, m$$

not all equal to zero, such that

$$\alpha(f(x) - \mu + \gamma_0) + \sum_{i=1}^{m} \beta_i(g_i(x) + \gamma_i) \geq 0$$

for all $x \in \mathcal{U}$ and all $\gamma_i > 0$, $i = 1, \ldots, m$. Suppose we pick any arbitrary $x \in \mathcal{U}$ and fix γ_i, $i = 1, \ldots, m$. As we let $\gamma_0 = k \to \infty$, we obtain

$$\alpha \geq O\left(\frac{1}{k}\right) \to 0.$$

We can repeat this process for each β_i, $i = 1, \ldots, m$ to conclude that

$$\beta_i \geq O\left(\frac{1}{k}\right) \to 0$$

for $i = 1, \ldots, m$ as $k \to \infty$. On the other hand, if we take the limit as

$$\gamma_i \to 0$$

for $i = 1, \ldots, m$, we obtain

$$\alpha(f(x) - \mu) + \sum_{i=1}^{m} \beta_i g_i(x) \geq 0.$$

The theorem is proved. $\qquad\square$

In essence, the only property that has been employed in the proof of this theorem has been the convexity of the functionals f, g_i, $i = 1, \ldots, m$ and the convexity of the constraint set $\mathcal{U}$. Unfortunately, if the scalar α turns out to be precisely zero, $\alpha \equiv 0$, the conclusions that can be drawn from the above theorem are not very profound. In this *degenerate case*, the theorem simply asserts a self-evident property of the constraint functions. Significant attention has been directed toward the study of conditions that guarantee that this degenerate case does not occur. If $\alpha > 0$, the multipliers appearing in the theorem are said to be *regular multipliers*. Perhaps the simplest condition guaranteeing that the multipliers are indeed regular, $\alpha > 0$, is the *Slater condition* embodied in the following theorem.

Theorem 6.3.2. *Let $f : \mathcal{U} \subseteq X \to \mathbb{R}$ and $g_i : \mathcal{U} \subseteq X \to \mathbb{R}$, $i = 1, \ldots, m$ be proper convex functionals over the convex set $\mathcal{U} \subseteq X$. Suppose that the functionals g_i, $i = 1, \ldots, m$ satisfy the Slater condition: there is some $\overline{x} \in \mathcal{U} \subseteq \mathbb{R}^n$ such that*

$$g_i(\overline{x}) < 0 \qquad \forall\, i = 1, \ldots, m.$$

Then there exist real numbers $\lambda_{0,i} \geq 0$, $i = 1, \ldots, m$ such that

$$\mu \leq f(x) + \sum_{i=1}^{m} \lambda_{0,i} g_i(x)$$

for all $x \in \mathcal{U}$. Moreover,

$$\mu = \inf\left\{ f(x) + \sum_{i=1}^{m} \lambda_{0,i} g_i(x) : x \in \mathcal{U} \right\}. \tag{6.10}$$

Proof. From the proof of the last theorem, we know that there are real numbers $\alpha \geq 0$, $\beta_i \geq 0$, $i = 1, \ldots, m$ such that

$$\alpha(f(x) - \mu) + \sum_{i=1}^{m} \beta_i g_i(x) \geq 0$$

for all $x \in \mathcal{U}$. Suppose to the contrary that the multipliers are degenerate, that is, $\alpha = 0$. Hence,

$$\sum_{i=1}^{m} \beta_i g_i(x) \geq 0. \tag{6.11}$$

By the Slater condition there is an $\bar{x}$ such that $g_i(\bar{x}) < 0$ for each $i = 1, \ldots, m$. In view of (6.11), the Slater condition implies that $\beta_i = 0$ for $i = 1, \ldots, m$. But this is a contradiction of the requirement that not all of the coefficients α, β_i, $i = 1, \ldots, m$ are zero. We conclude that $\alpha > 0$ when the Slater condition is fulfilled. Dividing by the coefficient α we obtain

$$\mu \leq f(x) + \sum_{i=1}^{m} \frac{\beta_i}{\alpha} g_i(x)$$

for all $x \in \mathcal{U}$ and the first assertion of the theorem is proved for the choice $\lambda_{0,i} \triangleq \beta_i/\alpha$, $i = 1, \ldots, m$. Thus, we have shown that

$$\mu \leq f(x) + \sum_{i=1}^{m} \lambda_{0,i} g_i(x)$$

for all $x \in \mathcal{U}$. It is also clear that

$$\mu \leq C_1 \triangleq \inf \left\{ f(x) + \sum_{i=1}^{m} \lambda_{0,i} g_i(x) : x \in \mathcal{U} \right\}.$$

If we restrict the set over which we take the infimum, the value of the infimum can only increase. We can deduce that

$$\mu \leq C_1 \leq C_2 \triangleq \inf \left\{ f(x) + \sum_{i=1}^{m} \lambda_{0,i} g_i(x) : x \in \mathcal{U}, \ g_i(x) \leq 0, \ i = 1, \ldots, m \right\}.$$

But since $g_i(x) \leq 0$ implies $\lambda_{0,i} g_i(x) \leq 0$, we conclude that

$$\mu \leq C_1 \leq C_2 \leq \mu \triangleq \inf \left\{ f(x) : g_i(x) \leq 0, \ i = 1, \ldots, m, \ x \in \mathcal{U} \right\}.$$

Hence,

$$\mu = \inf \left\{ f(x) + \sum_{i=1}^{m} \lambda_{0,i} g_i(x) \right\}. \qquad \square$$

The familiar characterization of constrained minima for convex functionals defined on convex subsets of $\mathbb{R}^n$ can now be stated in terms of Lagrange multipliers. Define the Lagrange functional

$$L(x, \lambda) \stackrel{\triangle}{=} f(x) + \sum_{i=1}^{m} \lambda_i g_i(x).$$

Theorem 6.3.3. *Let $f : \mathcal{U} \subseteq \mathbb{R}^n \to \mathbb{R}$ and $g_i : \mathcal{U} \subseteq \mathbb{R}^n \to \mathbb{R}$, $i = 1, \ldots, m$, be proper convex functions defined on the convex set $\mathcal{U} \subseteq \mathbb{R}^n$. Suppose that the functions g_i, $i = 1, \ldots, m$, satisfy the Slater condition*

$$g_i(\overline{x}) < 0 \qquad \forall i = 1, \ldots, m$$

for some $\overline{x} \in \mathcal{U}$. If there is a point $x_0 \in \mathbb{R}^n$ at which the infimum is attained in the minimization problem

$$f(x_0) = \mu = \inf \left\{ f(x) : g_i(x) \leq 0, \ x \in \mathbb{R}^n, \ i = 1, \ldots, m \right\}$$

and

$$g(x_0) \leq 0$$

then there are real numbers

$$\lambda_{0,i} \geq 0 \qquad \forall \, i = 1, \ldots, m,$$

such that we have

$$\sum_{i=1}^{m} \lambda_{0,i} g_i(x_0) = 0$$

and

$$L(x_0, \lambda) \leq L(x_0, \lambda_0) \leq L(x, \lambda_0)$$

for all $x \in \mathcal{U}$ and $\lambda \in [0, \infty)^m$.

Proof. First we will show that

$$\sum_{i=1}^{m} \lambda_{0,i} g_i(x_0) = 0.$$

We know from the last theorem that

$$
\begin{aligned}
\mu = f(x_0) \ &= \ \inf \left\{ f(x) + \sum_{i=1}^{m} \lambda_{0,i} g_i(x) \right\} \\
&\leq \ f(x_0) + \sum_{i=1}^{m} \lambda_{0,i} g_i(x_0)
\end{aligned}
$$

or that

$$0 \le \sum_{i=1}^{m} \lambda_{0,i} g_i(x_0).$$

On the other hand, since $\lambda_{0,i} \ge 0$ and $g_i(x_0) \le 0$, we know that

$$0 \ge \sum_{i=1}^{m} \lambda_{0,i} g_i(x_0).$$

Combining these results, we conclude that

$$\sum_{i=1}^{m} \lambda_{0,i} g_i(x_0) = 0.$$

Now we can show that the pair (x_0, λ_0) is a saddle point for the Lagrange functional. By definition, we have

$$
\begin{aligned}
L(x_0, \lambda) &= f(x_0) + \underbrace{\sum_{i=1}^{m} \lambda_i g_i(x_0)}_{\le 0} \\
&\le f(x_0) + \underbrace{\sum_{i=1}^{m} \lambda_{0,i} g_i(x_0)}_{=0} = L(x_0, \lambda_0) \\
&\le f(x) + \sum_{i=1}^{m} \lambda_{0,i} g_i(x) = L(x, \lambda_0).
\end{aligned}
$$

This last inequality follows from Equation (6.10). Hence, altogether we have

$$L(x_0, \lambda) \le L(x_0, \lambda_0) \le L(x, \lambda_0)$$

and the point (x_0, λ_0) is a saddle point for the Lagrange functional $L(x, \lambda)$. $\square$

6.4   Ordered Vector Spaces

In the last section, we discussed the role of Lagrange multipliers in the extremization of convex functionals defined in terms of real variables. In this section, we study the generalization of this approach to convex functionals defined over vector spaces that are endowed with an ordering. The constraints imposed via inequalities are couched in terms of the ordering, which is induced via the introduction of a positive cone.

6.4.1 Positive Cones, Negative Cones, and Orderings

We have already seen that convexity plays an important role in numerous optimization problems. In addition, some classes of inequality constraints couched in terms of convex cones are amenable to Lagrange multiplier methods for abstract spaces. In this section we discuss how convex cones can be used to define orderings on vector spaces, which in turn facilitate the introduction of inequality constraints.

Definition 6.4.1. *Let X be a vector space. A set $C_0 \subseteq X$ is a cone with vertex at the origin if*

$$x \in C_0 \quad implies \quad \alpha x \in C_0.$$

for all $\alpha \geq 0$. A cone C with vertex at a point $v_0 \in X$ is defined to be the translation $v_0 + C_0$ of a cone C_0 with vertex at the origin

$$C = \{\, v_0 + x \, : \, \forall\, x \in C_0 \,\},$$

Definition 6.4.2. *A cone C is pointed if it does not contain a line.*

Definition 6.4.3. *Let X be a vector space. A positive cone $C_0^+ \subseteq X$ is defined by*

$$C_0^+ = \{\, x \, : \, x \in C_0, \; x \geq 0 \,\}.$$

These definitions are widely employed. Examples of cones in $\mathbb{R}^2$ are depicted schematically in Figure 6.7. Some features of the definition should be noted.

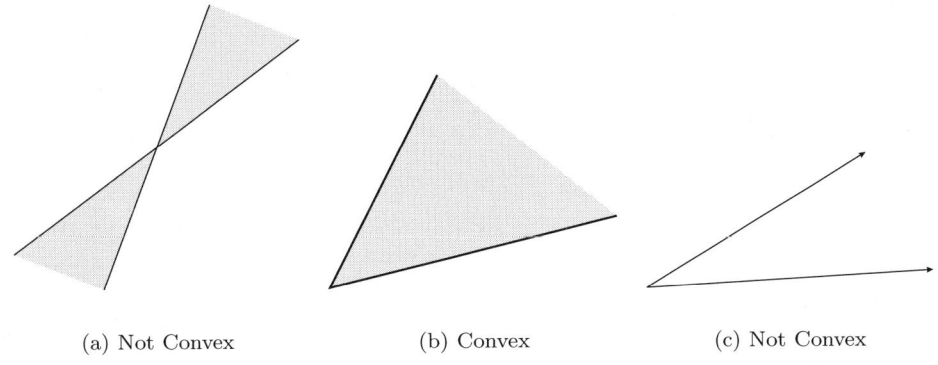

  (a) Not Convex   (b) Convex   (c) Not Convex

Figure 6.7: Examples of Cones in $\mathbb{R}^2$

A cone C with vertex at the origin need not be convex. Examples of cones that are not convex are depicted in Figures 6.7a and 6.7c. Just as importantly, a cone need not be closed, in general. For example, Figure 6.8a depicts the open cone obtained via translation of the open cone with vertex at origin

$$\{(x, y) \in \mathbb{R}^2 : x > 0, \, y > 0\}.$$

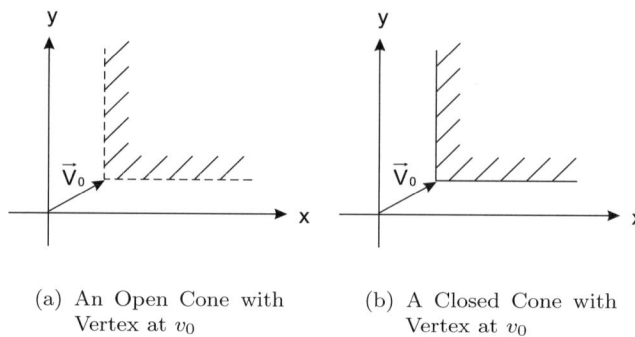

(a) An Open Cone with
    Vertex at v_0

(b) A Closed Cone with
    Vertex at v_0

Figure 6.8: Open and Closed Cones

Figure 6.8b depicts the closed cone obtained via translation of the closed cone with vertex at the origin

$$\{(x, y) \in \mathbb{R}^2 : x \geq 0,\ y \geq 0\}.$$

In our applications, we will be primarily concerned with closed convex cones. Unless we state otherwise, when we refer to a cone we will always mean a cone with vertex at the origin.

To define inequality constraints in extremization problems, it is important to note that convex cones induce orderings on a vector space.

Proposition 6.4.1. *Let C be a convex pointed cone with vertex at the origin contained in a vector space X. The relation R defined by*

$$uRv \quad \Longleftrightarrow \quad u - v \in C$$

defines an ordering on X.

Proof. Recall that an ordering is a relationship that is reflexive, transitive, and antisymmetric. The relation R is reflexive if uRu for all $u \in V$. But this is always true since $0 = u - u \in C$. A relationship R is transitive if

$$uRv \quad \text{and} \quad vRw \quad \Longrightarrow \quad uRw.$$

But suppose

$$u - v \in C \quad \text{and} \quad v - w \in C.$$

Since C is convex, we have

$$\tfrac{1}{2}(u - v) + \tfrac{1}{2}(v - w) \in C$$
$$\tfrac{1}{2}(u - w) \in C.$$

Since C is a cone, this implies that

$$u - w \in C$$

or, equivalently
$$uRw.$$
The relation is therefore transitive. Finally, the relation R is antisymmetric if
$$uRv \quad \text{and} \quad vRu \quad \Longrightarrow \quad u = v.$$
The left-hand side implies that
$$u - v \in C \quad \text{and} \quad -(u - v) \in C$$
which means that the cone C contains the line $\lambda(u - v)$, where $\lambda \in \mathbb{R}$. But since
the cone is pointed, we have $u - v = 0$. $\qquad\square$

A graphical representation of the ordering induced by a convex pointed cone
with vertex at the origin is given in Figure 6.9.

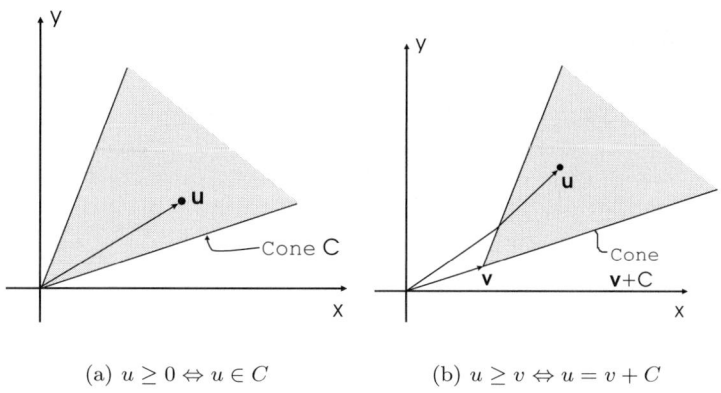

$$\text{(a) } u \geq 0 \Leftrightarrow u \in C \qquad\qquad \text{(b) } u \geq v \Leftrightarrow u = v + C$$

Figure 6.9: Ordering Associated with a Convex Cone

Note that to define an ordering on V it is not enough for a cone to have
vertex at the origin and be convex. Indeed, for any constant a, a half-space $au \geq 0$
is a convex cone with vertex at the origin. However, for any u_1 and u_2 such that
$au_1 = 0$ and $au_2 = 0$, an ordering is not defined since from $u_1 R u_2$ and $u_2 R u_1$ we
cannot conclude that $u_1 = u_2$.

6.4.2   Orderings on Sobolev Spaces

Some of the most important applications and examples of orderings, and their
associated cones, in this chapter are defined on Sobolev spaces. These orderings
will facilitate the introduction of constraints that represent contact between classes
of elastic continua. In this section, we introduce orderings on the Sobolev space
$H^1(\Omega)$. Extensions to more general Sobolev spaces follow similar reasoning.

Definition 6.4.4. *A function $u \in H^1(\Omega)$ is said to be nonnegative in the sense of $H^1(\Omega)$ if there exists a sequence $\{u_k\}_{k \in \mathbb{N}} \subseteq W^{1,\infty}(\Omega)$ such that*

$$u_k \to u \quad in \quad H^1(\Omega)$$

and

$$u_k(x) \geq 0$$

for all $x \in \Omega$.

Recall that $W^{1,\infty}(\Omega) \subseteq W^{1,2}(\Omega) \equiv H^1(\Omega)$.

More general definitions of nonnegativity in the Sobolev space $H^1(\Omega)$, given in terms of subsets of Ω, are possible. This definition will suffice for the examples considered in this chapter. The interested reader should see [21] or [13] for detailed discussions of these cases. We will abbreviate the phrase "u is nonnegative in the sense of $H^1(\Omega)$" as

$$u \gtrsim 0$$

in the sequel. The following proposition makes clear that the relation $\gtrsim$ is an ordering on $H^1(\Omega)$ and defines a closed convex cone.

Proposition 6.4.2. *The set*

$$C \overset{\triangle}{=} \{f \in H^1(\Omega) : \ f \gtrsim 0\}$$

is a closed convex cone with vertex at the origin in $H^1(\Omega)$.

Proof. First, we show that C is a closed set in $H^1(\Omega)$. Suppose that $\{u_k\}_{k \in \mathbb{N}} \subseteq C$ and that

$$u_k \to u_0 \quad in \quad H^1(\Omega).$$

By the definition of the set C for each fixed u_k there is a sequence $\{u_{k,j}\}_{j \in \mathbb{N}} \subseteq W^{1,\infty}(\Omega)$ such that

$$u_{k,j} \to u_k \quad in \quad H^1(\Omega)$$

as $j \to \infty$ and for each $j \in \mathbb{N}$

$$u_{k,j}(x) \geq 0$$

for all $x \in \Omega$. But in this case the doubly-indexed sequence $\{u_{k,j}\}_{k,j \in \mathbb{N}} \subseteq W^{1,\infty}(\Omega)$ satisfies

$$u_{k,j} \to u_0 \quad in \quad H^1(\Omega)$$

as $k, j \to \infty$ and for each pair $(k, j) \in \mathbb{N} \times \mathbb{N}$

$$u_{k,j}(x) \geq 0.$$

That $\{u_{k,j}\}_{k,j \in \mathbb{N}}$ converges to u_0 follows since

$$\|u_0 - u_{k,j}\|_{H^1(\Omega)} \leq \|u_0 - u_k\|_{H^1(\Omega)} + \|u_k - u_{k,j}\|_{H^1(\Omega)} \to 0$$

as $k, j \to \infty$. But this simply means that $u_0 \in C$, and the set C is consequently closed in $H^1(\Omega)$. Moreover, C is clearly a cone in $H^1(\Omega)$. Suppose $u_0 \in C$. By definition there exists a sequence $\{u_k\}_{k \in \mathbb{N}} \subseteq W^{1,\infty}(\Omega)$ such that

$$u_k \to u_0 \in H^1(\Omega)$$

and for each $k \in \mathbb{N}$

$$u_k(x) \geq 0$$

for all $x \in \Omega$. But for any $\alpha \geq 0$, the sequence $\{\alpha u_k\}_{k \in \mathbb{N}} \subseteq W^{1,\infty}(\Omega)$ and

$$\|\alpha u_k - \alpha u_0\|_{H^1(\Omega)} = |\alpha| \|u_k - u_0\|_{H^1(\Omega)} \to 0$$

as $k \to \infty$ and for each k

$$\alpha u_k(x) \geq 0$$

for all $x \in \Omega$. In other words, $\alpha u \in C$ and C is a cone with vertex at the origin. Finally, the cone C is convex. Suppose $u_0, v_0 \in C$, with associated sequences $\{u_k\}_{k \in \mathbb{N}}, \{v_k\}_{k \in \mathbb{N}} \subseteq W^{1,\infty}(\Omega)$ satisfying

$$u_k \to u_0 \quad \text{in} \quad H^1(\Omega)$$
$$v_k \to v_0 \quad \text{in} \quad H^1(\Omega)$$

and for each $k \in \mathbb{N}$

$$u_k(x) \geq 0$$
$$v_k(x) \geq 0$$

for all $x \in \Omega$. Then $\lambda u_0 + (1 - \lambda)v_0 \in C$ for all $\lambda \in [0, 1]$ since the sequence $\{\lambda u_k + (1 - \lambda)v_k\}_{k \in \mathbb{N}} \subseteq W^{1,\infty}(\Omega)$

$$\lambda u_k + (1 - \lambda)v_k \to \lambda u_0 + (1 - \lambda)v_0$$

in $H^1(\Omega)$ and for each k

$$\lambda u_k(x) + (1 - \lambda)v_k(x) \geq 0$$

for all $x \in \Omega$. $\qquad \square$

6.5 Convex Programming in Ordered Vector Spaces

Suppose that $f : X \to \mathbb{R}$, $g : X \to Y$, X and Y are vector spaces, and $\mathcal{U}$ is a convex set contained in X. In this section, we study the optimization problem in which we seek $x_0 \in \mathcal{U} \subseteq X$ such that

$$f(x_0) = \mu = \inf \{f(x) : g(x) \leq 0, \ x \in \mathcal{U} \subseteq X\}$$

and

$$g(x_0) \leq 0.$$

Suppose that the vector space Y contains a cone P, that induces an ordering of Y. Hence, expressions like

$$g(x) \leq 0$$

are understood in the sense discussed in Section 6.4.1. That is, if y_1, $y_2 \in Y$

$$y_2 \geq y_1 \quad \Longleftrightarrow \quad y_2 - y_1 \in P.$$

Further, we will need a notion of dual cone.

Definition 6.5.1. *The cone P^* dual to P, or dual cone P^*, is defined by*

$$P^* = \{h \in Y^* : \langle h, y \rangle_{Y^* \times Y} \geq 0 \ \forall y \in P\}.$$

As in the case of functionals defined over the real variables, the existence of a Lagrange multiplier depends in a critical fashion on the application of the Hahn-Banach theorem.

Theorem 6.5.1. *Let $f : X \to \mathbb{R}$ and $g : X \to Y$ be proper convex functions defined over the convex set $\mathcal{U} \subseteq X$. Let P be a positive cone in Y and suppose P has an interior point. Then there exists a pair $(\alpha, \beta^*) \in \mathbb{R} \times Y^*$ such that*

$$\alpha(f(x) - \mu) + \langle \beta^*, g(x) \rangle \geq 0$$

where

$$\alpha \geq 0, \qquad \beta^* \geq 0.$$

The notation $\beta^* \geq 0$ means that $\beta^* \in P^*$.

Proof. The proof of this theorem, again, relies in a fundamental way on the Hahn-Banach theorem. Define the sets

$$C_1 \stackrel{\triangle}{=} \{(\xi, \eta) : \xi \geq f(x), \ \eta \geq g(x), \ \text{for some } x \in \mathcal{U}\}$$

$$C_2 \stackrel{\triangle}{=} \{(\xi, \eta) : \xi \leq \mu, \ \eta \leq 0\}.$$

The sets C_1 and C_2 are convex. Suppose that (ξ_1, η_1) and (ξ_2, η_2) are elements of C_1. By the definition of C_1, there exist x_1 and x_2 in $\mathcal{U}$ such that

$$\lambda \xi_1 + (1 - \lambda)\xi_2 \geq \lambda f(x_1) + (1 - \lambda)f(x_2)$$

and by the convexity of f it follows that

$$\lambda \xi_1 + (1 - \lambda)\xi_2 \geq f(\lambda x_1 + (1 - \lambda)x_2). \tag{6.12}$$

Similarly, we have

$$\begin{aligned} \lambda \eta_1 + (1 - \lambda)\eta_2 &\geq \lambda g(x_1) + (1 - \lambda)g(x_2) \\ &\geq g(\lambda x_1 + (1 - \lambda)x_2). \end{aligned} \tag{6.13}$$

Note that we have used the ordering property

$$a \leq b \quad \text{and} \quad c \leq d \quad \Longleftrightarrow \quad a + c \leq b + d$$

discussed in the introduction to this chapter. From the convexity of $\mathcal{U}$, we know that $\lambda x_1 + (1 - \lambda)x_2 \in \mathcal{U}$ for $\lambda \in [0, 1]$. Consequently, the inequalities (6.12) and (6.13) imply that

$$\lambda(\xi_1, \eta_1) + (1 - \lambda)(\xi_2, \eta_2) \in C_1$$

corresponding to the point

$$\lambda x_1 + (1 - \lambda)x_2 \in \mathcal{U}.$$

In a like manner, if (ξ_1, η_1) and (ξ_2, η_2) are contained in C_2, it follows that

$$\begin{aligned} \lambda\xi_1 + (1 - \lambda)\xi_2 &\leq \lambda\mu + (1 - \lambda)\mu = \mu \\ \lambda\eta_1 + (1 - \lambda)\eta_2 &\leq \lambda \cdot 0 + (1 - \lambda) \cdot 0 = 0. \end{aligned}$$

Taken together, these last two inequalities imply that

$$(\xi_1, \eta_1), \ (\xi_2, \eta_2) \in C_2 \quad \Longrightarrow \quad \lambda(\xi_1, \eta_1) + (1 - \lambda)(\xi_2, \eta_2) \in C_2.$$

Sets C_1 and C_2 are not only convex, but C_1 also contains no interior points of C_2. If the point $(\xi, \eta) \in \text{int}(C_2)$, then by definition we have

$$\xi \quad < \quad \mu \tag{6.14}$$

$$\eta \quad < \quad 0. \tag{6.15}$$

In the latter case, the condition that $\eta < 0$ is equivalent to

$$\begin{aligned} 0 - \eta &\in \text{int}(P) \\ \eta &\in \text{int}(-P). \end{aligned}$$

On the other hand, if the point $(\xi, \eta) \in C_1$, then there is some $\hat{x} \in \mathcal{U}$ such that

$$\xi \geq f(\hat{x}) \quad \text{and} \quad \eta \geq g(\hat{x}). \tag{6.16}$$

Thus we have, by combining inequalities (6.15), (6.14) and (6.16), that there is some $\hat{x} \in \mathcal{U}$ such that

$$0 > \eta \geq g(\hat{x})$$

and

$$\xi \geq f(\hat{x}) \geq \inf\{f(x) : g(x) \leq 0, \ x \in \mathcal{U}\} = \mu.$$

But this inequality contradicts the fact that $(\xi, \eta) \in \text{int}(C_2)$ as characterized in (6.14), and we conclude that $C_1 \cap \text{int}(C_2) = \emptyset$. In addition, it is not difficult to argue that the set C_2 contains an interior point. By definition,

$$(\xi, \eta) \in \text{int}(C_2)$$

if and only if

$$\xi < \mu \quad \text{and} \quad \eta < 0.$$

Since f is proper, μ is finite. By hypothesis the positive cone P contains an interior point. It follows, therefore, that the set C_2 contains an interior point. By Eidelheit's separation theorem, a variant of the Hahn-Banach theorem, for all $(\xi_1, \eta_1) \in C_1$ and $(\xi_2, \eta_2) \in C_2$, there exists a linear functional $(\alpha, \beta^*) \in \mathbb{R} \times Y^*$ such that

$$\alpha(\xi_1 - \xi_2) + \langle \beta^*, \eta_1 - \eta_2 \rangle_{Y^* \times Y} \geq 0. \tag{6.17}$$

We claim that, in fact, we have

$$\alpha \geq 0, \qquad \beta^* \geq 0.$$

Recall that

$$\mu = \inf \{ f(x) : g(x) \leq 0, \ x \in \mathcal{U} \subseteq X \}.$$

By the definition of the infimum, there is a sequence $\{x_k\}_{k \in \mathbb{N}} \subset \mathcal{U}$ such that

$$\begin{aligned} f(x_k) &\rightarrow \mu, & f(x_k) &\geq \mu, & k &\rightarrow \infty \\ g(x_k) &\leq 0 & \forall k &\in \mathbb{N}. \end{aligned}$$

Then the sequence $\{(f(x_k), 0)\}_{k \in \mathbb{N}} \subset C_1$ and $(\mu, 0) \in C_2$. Substituting these values in the separation inequality (6.17), we obtain

$$\alpha \underbrace{(f(x_k) - \mu)}_{\geq 0} \geq 0$$

and consequently $\alpha \geq 0$. In addition, for any $(f(x_k), 0) \in C_1$ and $(\xi, \eta) \in C_2$, we have

$$\alpha(f(x_k) - \xi) + \langle \beta^*, 0 - \eta \rangle_{Y^* \times Y} \geq 0.$$

By the definition of C_2,

$$\eta \leq 0$$

which is equivalent to

$$-\eta \in P.$$

By the definition of the positive cone,

$$(-\eta) \in P \quad \Longleftrightarrow \quad n(-\eta) \in P$$

for every positive integer $n \in \mathbb{N}$. When we evaluate the separation inequality at $-n\eta \in P$, we obtain

$$\langle \beta^*, -n\eta \rangle_{Y^* \times Y} \geq -\alpha(f(x_k) - \xi)$$

or

$$\langle \beta^*, -\eta \rangle_{Y^* \times Y} \geq -\frac{\alpha}{n}(f(x_k) - \xi)$$

$$\langle \beta^*, -\eta \rangle_{Y^* \times Y} \geq -O\left(\frac{1}{n}\right) \rightarrow 0$$

which approaches zero as $n \rightarrow \infty$. Since $\eta \in -P$, we have $-\eta \in P$ and, therefore,

$$\beta^* \in P^* \quad \Longleftrightarrow \quad \beta^* \geq 0. \qquad \qquad \Box$$

As in the last section, the key property employed to establish the existence of the Lagrange multipliers has been the convexity of f, g and the constraint set $\mathcal{U}$. Again, the conclusion of this theorem is not very strong in the *degenerate* case when $\alpha \equiv 0$. The generalization of the Slater condition to the current framework is straightforward, and guarantees that the Lagrange multipliers are *regular*.

Theorem 6.5.2. *Let $f : X \to \mathbb{R}$ and $g : X \to Y$ be proper convex functions defined on the convex set $\mathcal{U} \subset X$. Let P be a positive cone in Y that contains an interior point. Suppose that the function g satisfies the Slater condition*

$$g(\overline{x}) < 0$$

for some $\overline{x} \in \mathcal{U}$. Then there exists a $\lambda_0^* \geq 0$ such that

$$\mu \leq f(x) + \langle \lambda_0^*, g(x) \rangle_{Y^* \times Y}$$

for all $x \in \mathcal{U}$. Moreover,

$$\mu = \inf \left\{ f(x) + \langle \lambda_0^*, g(x) \rangle_{Y^* \times Y} : x \in \mathcal{U} \right\}.$$

Proof. From the last theorem, we know that there exists a pair $(\alpha, \beta^*) \in \mathbb{R} \times Y^*$ such that

$$\alpha(\xi_1 - \xi_2) + \langle \beta^*, \eta_1 - \eta_2 \rangle_{Y^* \times Y} \geq 0$$

for all $(\xi_1, \eta_1) \in C_1$ and $(\xi_2, \eta_2) \in C_2$, where $\alpha \geq 0$ and $\beta^* \geq 0$. Since $(\mu, 0) \in C_2$, we have in particular that

$$\alpha(\xi_1 - \mu) + \langle \beta^*, \eta_1 \rangle_{Y^* \times Y} \geq 0$$

for all $(\xi_1, \eta_1) \in C_1$. Suppose that $\alpha \equiv 0$, which is the degenerate case. Then,

$$\langle \beta^*, \eta_1 \rangle_{Y^* \times Y} \geq 0$$

for all η_1 such that

$$\eta_1 \geq g(x)$$

for some $x \in \mathcal{U}$. For the choice $x = \overline{x}$ defined in Slater condition, we consider $\eta_1 = g(\overline{x})$ and obtain

$$\langle \beta^*, g(\overline{x}) \rangle_{Y^* \times Y} \geq 0.$$

Since $\beta^* \geq 0$, this contradicts the hypothesis that $g(\overline{x}) < 0$, i.e.,

$$\langle \beta^*, g(\overline{x}) \rangle_{Y^* \times Y} < 0.$$

We conclude that $\alpha > 0$. Since $\alpha > 0$, there exists a $\lambda_0^* \geq 0$

$$\lambda_0^* = \frac{1}{\alpha} \beta^*$$

such that

$$\mu \leq \xi_1 + \langle \lambda_0^*, \eta_1 \rangle_{Y^* \times Y}$$

for all $(\xi_1, \eta_1) \in C_1$. Clearly, we have

$$
\begin{aligned}
\mu &\leq \inf\{\xi + \langle \lambda_0^*, \eta \rangle_{Y^* \times Y} : (\xi, \eta) \in C_1\} \\
&\leq \{\xi + \langle \lambda_0^*, \eta \rangle_{Y^* \times Y} : (\xi, \eta) = (f(x), g(x)), \ x \in \mathcal{U}\} \\
&= \inf\{f(x) + \langle \lambda_0^*, g(x) \rangle_{Y^* \times Y} : x \in \mathcal{U}\}.
\end{aligned}
$$

However, since $\langle \lambda_0^*, g(x) \rangle_{Y^* \times Y} \leq 0$ then

$$
\begin{aligned}
\inf\{f(x) + \langle \lambda_0^*, g(x) \rangle_{Y^* \times Y} : x \in \mathcal{U}\} &\leq \inf\{f(x) : x \in \mathcal{U}\} \\
&\leq \inf\{f(x) : g(x) \leq 0, \ x \in \mathcal{U}\} \\
&= \mu.
\end{aligned}
$$

As a result of the last two inequalities, we obtain

$$
\mu = \inf\{f(x) + \langle \lambda_0^*, g(x) \rangle_{Y^* \times Y} : x \in \mathcal{U}\}. \qquad \square
$$

As in the case when we studied functionals defined in terms of real variables, the current case can also be cast in terms of a Lagrange functional

$$
\begin{aligned}
L &: \quad X \times Y^* \to \mathbb{R} \\
L(x, \lambda^*) &\overset{\triangle}{=} f(x) + \langle \lambda^*, g(x) \rangle_{Y^* \times Y}.
\end{aligned}
$$

Theorem 6.5.3. *Let $f : X \to \mathbb{R}$ and $g : X \to Y$ be proper convex functions defined on the convex set $\mathcal{U} \subset X$. Let P be a positive cone in Y that contains an interior point. Suppose that the function g satisfies the Slater condition*

$$
g(\overline{x}) < 0
$$

for some $\overline{x} \in \mathcal{U}$. If there is an $x_0 \in \mathcal{U}$ such that

$$
f(x_0) = \mu = \inf\{f(x) : g(x) \leq 0, \ x \in \mathcal{U} \subset X\}
$$

and

$$
g(x_0) \leq 0
$$

then there is a $\lambda_0^* \geq 0$ such that

$$
\langle \lambda_0^*, g(x_0) \rangle_{Y^* \times Y} = 0
$$

and

$$
L(x_0, \lambda^*) \leq L(x_0, \lambda_0^*) \leq L(x, \lambda_0^*)
$$

for all $x \in \mathcal{U}$ and $\lambda^* \in Y^*$.

Proof. From the last theorem, there is a $\lambda_0^* \geq 0$ such that

$$
\mu \leq f(x) + \langle \lambda_0^*, g(x) \rangle_{Y^* \times Y}
$$

for all $x \in \mathcal{U}$. Since x_0 is the point at which f attains the infimum

$$f(x_0) = \mu \leq f(x_0) + \langle \lambda_0^*, g(x_0) \rangle_{Y^* \times Y}$$

we obtain

$$0 \leq \langle \lambda_0^*, g(x_0) \rangle_{Y^* \times Y}.$$

Since $\lambda_0^* \geq 0$ and $g(x_0) \leq 0$, we have

$$\langle \lambda_0^*, g(x_0) \rangle_{Y^* \times Y} \leq 0.$$

Taken together, these inequalities imply that

$$\langle \lambda_0^*, g(x_0) \rangle_{Y^* \times Y} = 0.$$

Finally, we can directly compute

$$\begin{aligned}
L(x_0, \lambda^*) &= f(x_0) + \underbrace{\langle \lambda^*, g(x_0) \rangle_{Y^* \times Y}}_{\leq 0} \\
&\leq f(x_0) + \underbrace{\langle \lambda_0^*, g(x_0) \rangle_{Y^* \times Y}}_{=0} = L(x_0, \lambda_0^*).
\end{aligned}$$

Likewise, we have

$$\mu = f(x_0) + \underbrace{\langle \lambda_0^*, g(x_0) \rangle_{Y^* \times Y}}_{=0} \leq f(x) + \langle \lambda_0^*, g(x) \rangle_{Y^* \times Y}$$

$$= L(x, \lambda_0^*).$$

Combining these inequalities, we have

$$L(x_0, \lambda^*) \leq L(x_0, \lambda_0^*) \leq L(x, \lambda_0^*)$$

for all $x \in \mathrm{dom}(f) \cap \mathrm{dom}(g)$, and the theorem is proved. $\square$

6.6   Gateaux Differentiable Functionals on Ordered Vector Spaces

In this section we consider problems that have a structure quite similar to that considered in Section 6.4. Suppose $f : \mathcal{U} \subseteq X \to \mathbb{R}$ and $g : \mathcal{U} \subseteq X \to Y$, where X and Y are normed vector spaces and $\mathcal{U}$ is a convex subset of X. We study the optimization problem in which we seek $x_0 \in \mathcal{U} \subseteq X$ such that

$$f(x_0) = \inf\{f(x) : g(x) \leq 0, \ x \in \mathcal{U}\}$$

and

$$g(x_0) \leq 0.$$

The function g embodies the constraints on the optimization problem, and the normed vector space Y contains a cone P, that induces an ordering of Y. Hence, expressions like

$$g(x) \leq 0$$

are understood in the sense discussed in Section 6.4.1. That is, if $y_1, y_2 \in Y$,

$$y_2 \geq y_1 \quad \Longleftrightarrow \quad y_2 - y_1 \in P.$$

Up to this point, we have not focussed on techniques to characterize the minimizers of such problems. The following theorem is prototypical of multiplier methods that do characterize these solutions.

Theorem 6.6.1. *Suppose that $\mathcal{U}$ is convex, $f : X \rightarrow \mathbb{R}$ is Gateaux differentiable, $g : X \rightarrow Y$ is Gateaux differentiable, and P has nonempty interior. Then every solution x_0 to the minimization problem satisfies*

$$\left(Df(x_0)\right)' \circ \lambda_0 + \left(Dg(x_0)\right)' \circ \lambda_1 \in -N_{\mathcal{U}}(x_0) \tag{6.18}$$

$$\langle \lambda_1, g(x_0) \rangle_{Y^* \times Y} = 0 \tag{6.19}$$

$$\lambda_0 + \|\lambda_1\|_{Y^*} > 0 \tag{6.20}$$

for some $\lambda_0 \in \mathbb{R}$ and $\lambda_1 \in Y^*$ such that $\lambda_0 \geq 0$ and $\lambda_1 \geq 0$, where the normal cone is defined by

$$N_{\mathcal{U}}(x_0) = \{\xi \in X^* : \langle \xi, x_0 - x \rangle_{X^* \times X} \geq 0 \quad \forall x \in \mathcal{U}\}.$$

Proof. Before proving the above assertion, let us carefully interpret each term comprising Equations (6.18), (6.19) and (6.20). Since $f : X \rightarrow \mathbb{R}$, the Gateaux differential of f is just

$$Df(x_0, v) = \lim_{\epsilon \to 0} \frac{f(x_0 + \epsilon v) - f(x_0)}{\epsilon} \qquad \forall v \in X$$

so that

$$\langle Df(x_0), v \rangle_{X^* \times X} = Df(x_0, v) \qquad \forall v \in X$$

and, consequently $Df(x_0) \in X^*$. Similarly, $g : X \rightarrow Y$, the Gateaux differential is given by

$$Dg(x_0, v) = \lim_{\epsilon \to 0} \frac{g(x_0 + \epsilon v) - g(x_0)}{\epsilon} \qquad \forall v \in X.$$

We can rewrite these Gateaux derivatives

$$Df(x_0) \circ v = Df(x_0, v) \qquad \text{and} \qquad Dg(x_0) \circ v = Dg(x_0, v)$$

where $Df(x_0) \in \mathcal{L}(X, \mathbb{R})$ and $Dg(x_0) \in \mathcal{L}(X, Y)$.

We can define the transpose of these operators in the usual way. Recall that the transpose A' of the linear operator $A \in \mathcal{L}(X, Y)$ satisfies

$$\langle y^*, Ax \rangle_{Y^* \times Y} = \langle A'y^*, x \rangle_{X^* \times X}$$

$$A': Y^* \to X^*.$$

Therefore, we have

$$\left(Df(x_0)\right)': \mathbb{R} \to X^*$$

$$\langle \alpha, Df(x_0) \circ v \rangle_{\mathbb{R}^* \times \mathbb{R}} = \langle \left(Df(x_0)\right)' \circ \alpha, v \rangle_{X^* \times X}$$

and

$$\left(Dg(x_0)\right)': Y^* \to X^*$$

$$\langle y^*, Dg(x_0) \circ v \rangle_{Y^* \times Y} = \langle \left(Dg(x_0)\right)' \circ y^*, v \rangle_{X^* \times X}.$$

Now, we review the compositions that appear in Equation (6.18). Since we have

$$-\left[\left(Df(x_0)\right)' \circ \lambda_0 + \left(Dg(x_0)\right)' \circ \lambda_1 \right] \in N_{\mathcal{U}}(x_0)$$

the definition of the normal cone yields

$$-\langle \left(Df(x_0)\right)' \circ \lambda_0 + \left(Dg(x_0)\right)' \circ \lambda_1, x_0 - x \rangle_{X^* \times X} \geq 0 \quad \forall x \in \mathcal{U}.$$

By linearity, we have

$$\langle \left(Df(x_0)\right)' \circ \lambda_0, x - x_0 \rangle_{X^* \times X} + \langle \left(Dg(x_0)\right)' \circ \lambda_1, x - x_0 \rangle_{X^* \times X} \geq 0 \quad \forall x \in \mathcal{U}. \quad (6.21)$$

Moreover, for any arbitrary bounded linear operator A that is written as the product of two other bounded linear operators B and C

$$A = BC \qquad (6.22)$$

the transpose operator is given by

$$A' = C'B'. \qquad (6.23)$$

The equivalent representation for the Equation (6.18) is a consequence of Equations (6.21)–(6.23):

$$\langle \lambda_0, Df(x_0) \circ (x - x_0) \rangle_{\mathbb{R}^* \times \mathbb{R}} + \langle \lambda_1, Dg(x_0) \circ (x - x_0) \rangle_{Y^* \times Y} \geq 0. \qquad (6.24)$$

We conclude that the theorem follows if we can establish that Equations (6.24), (6.19) and (6.20) hold.

It is straightforward to establish that the following two subsets of $\mathbb{R} \times Y$

$$C_1 \triangleq \{(\xi, \eta) \in \mathbb{R} \times Y : \quad \xi \geq Df(x_0) \circ (x - x_0),$$
$$\eta \geq g(x_0) + Dg(x_0) \circ (x - x_0), \text{ for some } x \in \mathcal{U}\}$$

$$C_2 \triangleq \{(\xi, \eta) \in \mathbb{R} \times Y : \xi \leq 0, \ \eta \leq 0\}$$

are non-empty and convex. We show that $C_1 \cap \mathrm{int}(C_2) = \emptyset$. Assume to the contrary that $\exists (\xi, \eta) \in C_1 \cap \mathrm{int}(C_2)$. This fact implies that there is some $\tilde{x} \in \mathcal{U}$ such that

$$Df(x_0) \circ (\tilde{x} - x_0) < 0$$

$$g(x_0) + Dg(x_0) \circ (\tilde{x} - x_0) < 0. \tag{6.25}$$

By the definition of the Gateaux derivative, we can write

$$Df(x_0) \circ (\tilde{x} - x_0) = \lim_{\epsilon \to 0} \frac{f\big(x_0 + \epsilon(\tilde{x} - x_0)\big) - f(x_0)}{\epsilon} < 0.$$

Since this limit exists, it follows that $\exists \epsilon > 0$ such that

$$f\big(x_0 + \epsilon(\tilde{x} - x_0)\big) - f(x_0) < 0.$$

The inequality (6.25) implies that

$$0 - (g(x_0) + Dg(x_0) \circ (\tilde{x} - x_0)) \in \mathrm{int}(P).$$

By the definition of the interior of a set, there is an open ball centered at

$$\mathcal{B}_r(g(x_0) + Dg(x_0) \circ (\tilde{x} - x_0)) \in -P$$

for some $r > 0$. Since P is a cone, the dilation of this ball by a factor $0 < \gamma < 1$ is likewise contained in $-P$

$$\gamma \big[\mathcal{B}_r(g(x_0) + Dg(x_0) \circ (\tilde{x} - x_0)) \big] \in -P.$$

Since $-P$ is convex and $g(x_0) \leq 0$, i.e., $g(x_0) \in -P$,

$$\gamma(g(x_0) + Dg(x_0) \circ (\tilde{x} - x_0)) + (1 - \gamma)g(x_0) \in -P.$$

Simplifying, we have

$$g(x_0) + \gamma Dg(x_0) \circ (\tilde{x} - x_0) \in -P.$$

But this point can be selected arbitrarily close to $g(x_0 + \gamma(\tilde{x} - x_0))$ since

$$\|g(x_0 + \gamma(\tilde{x} - x_0)) - (g(x_0) + \gamma Dg(x_0) \circ (\tilde{x} - x_0))\|_Y$$
$$= \gamma \left\| \frac{g(x_0 + \gamma(\tilde{x} - x_0)) - g(x_0)}{\gamma} - Dg(x_0) \circ (\tilde{x} - x_0) \right\|_Y \to 0$$

as $\gamma \to 0$. Consequently, we can choose γ small enough so that $g(x_0 + \gamma(\tilde{x} - x_0)) \in -P$. In summary, we have shown that

$$f(x_0 + \gamma(\tilde{x} - x_0)) < f(x_0)$$

$$g(x_0 + \gamma(\tilde{x} - x_0)) \leq 0.$$

It means that point $x_0 + \gamma(\tilde{x} - x_0)$ is feasible and yields a value of f less than $f(x_0)$. This fact violates the optimality of x_0, and it follows that

$$C_1 \cap \text{int}(C_2) = \emptyset.$$

The sets C_1 and C_2 are convex and C_1 does not meet the interior of C_2. By Eidelheit's separation theorem [17, p. 133, Theorem 3] there is a closed hyperplane separating C_1 and C_2. The sets C_1 and C_2 are subsets of $\mathbb{R} \times Y$, so the hyperplane can be represented by a pair of functionals (λ_0, λ_1)

$$(\lambda_0, \lambda_1) \in (\mathbb{R} \times Y)^* = \mathbb{R} \times Y^*.$$

Since (λ_0, λ_1) separates C_1 and C_2, we have, in particular

$$(\lambda_0, \lambda_1) \circ (\xi, \eta) \geq c \qquad \forall (\xi, \eta) \in C_1$$
$$(\lambda_0, \lambda_1) \circ (\xi, \eta) \leq c \qquad \forall (\xi, \eta) \in C_2$$

for some constant c. In other words,

$$\lambda_0 \xi + \langle \lambda_1, \eta \rangle_{Y^* \times Y} \geq c \qquad \forall (\xi, \eta) \in C_1$$
$$\lambda_0 \xi + \langle \lambda_1, \eta \rangle_{Y^* \times Y} \leq c \qquad \forall (\xi, \eta) \in C_2. \tag{6.26}$$

In fact, $(0,0) \in C_1 \cap C_2$, and we conclude that $c \equiv 0$. From the second relation of (6.26), for any $(\xi, 0) \in C_2$, we obtain

$$\lambda_0 \, \xi \leq 0$$

but since $\xi \leq 0$, we have

$$\lambda_0 \geq 0.$$

Similarly, for any $(0, \eta) \in C_2$

$$\langle \lambda_1, \eta \rangle_{Y^* \times Y} \leq 0,$$

where, by the definition of C_2, $\eta \leq 0$. Consequently,

$$\lambda_1 \geq 0.$$

From the first relation of (6.26) and the definition of the set C_1, we obtain

$$\lambda_0 \, Df(x_0) \circ (x - x_0) + \langle \lambda_1, g(x_0) + Dg(x_0) \circ (x - x_0) \rangle_{Y^* \times Y} \geq 0. \tag{6.27}$$

for all $x \in \mathcal{U}$. For the choice $x = x_0$,

$$\langle \lambda_1, g(x_0) \rangle_{Y^* \times Y} \geq 0. \tag{6.28}$$

Relations $\lambda_1 \geq 0$ and $g(x_0) \leq 0$ imply that $\lambda_1 \in P^*$ and $g(x_0) \in -P$. Hence, we have

$$-\langle \lambda_1, g(x_0) \rangle_{Y^* \times Y} \geq 0. \tag{6.29}$$

From Equations (6.28) and (6.29), we obtain

$$\langle \lambda_1, g(x_0) \rangle_{Y^* \times Y} = 0$$

and Equation (6.27) becomes

$$\lambda_0 \, Df(x_0) \circ (x - x_0) + \langle \lambda_1, Dg(x_0) \circ (x - x_0) \rangle_{Y^* \times Y} \geq 0$$

$$\langle (Df(x_0))' \circ \lambda_0, (x - x_0) \rangle_{X^* \times X} + \langle (Dg(x_0))' \circ \lambda_1, (x - x_0) \rangle_{X^* \times X} \geq 0$$

for all $x \in \mathcal{U}$. Consequently, we have shown that Equations (6.18) and (6.19) hold. But since $(\lambda_0, \lambda_1) \neq (0, 0)$ in construction of the separating hyperplane, the relations $\lambda_0 \geq 0$ implies inequality (6.20) and the proof is complete. $\square$

Chapter 7

Lower Semicontinuous Functionals

Several important results, including the Weierstrass Theorem, may be established under weaker conditions than functional continuity. One such condition is known as lower semicontinuity, which has several equivalent definitions.

Definition 7.0.1. *Let (X, τ) be topological space and let $f : X \to \overline{\mathbb{R}}$. The functional f is lower semicontinuous at x_0 if the inverse image of every half-open set of the form (r, ∞), with $f(x_0) \in (r, \infty)$ contains an open set $U \subseteq X$ that contains x_0. That is,*

$$f(x_0) \in (r, \infty) \quad \Longrightarrow \quad \exists U \in \tau : x_0 \in U \subseteq f^{-1}(r, \infty)$$

As in the case of continuity, a function f is lower semicontinuous on a topological space X if it is lower semicontinuous at each point in X.

7.1 Characterization of Lower Semicontinuity

The next theorem establishes some alternative characterizations of *lower semicontinuity*.

Theorem 7.1.1. *Let (X, τ) be a topological space and let $f : X \to \overline{\mathbb{R}}$. The following are equivalent:*

(i) *The functional f is lower semicontinuous on X,*

(ii) *the inverse image of every half-open set of the form (r, ∞) is an open set in X, and*

(iii) *the inverse image of every half-closed set of the form $(-\infty, r]$ is a closed set in X.*

As a first step, let us define a version of the definition of lower semicontinuous functionals that is more appropriate for applications. Many applications are formulated in spaces with considerably more structure than a general topological space. The following proposition, couched in terms of a metric space, will suffice for a wide class of applications.

Proposition 7.1.1. *Let (X, d) be a metric space. A functional $f : X \to \overline{\mathbb{R}}$ is lower semicontinuous at a point $x_0 \in X$ if*

$$f(x_0) \leq \liminf_{x \to x_0} f(x).$$

Proof. Suppose f is lower semicontinuous at $x_0 \in X$. By definition, we know that

$$\liminf_{x \to x_0} f(x) \equiv \sup_{r>0} \inf_{x \in B(x_0, r)} f(x).$$

Now, from our earlier definition of lower semicontinuity, for each $\epsilon > 0$, there exists $r(\epsilon)$ such that

$$f(x_0) - \epsilon < f(x) \quad \text{for all} \quad x \in B(x_0, r(\epsilon))$$

or,

$$f(x_0) < \epsilon + \inf_{x \in B(x_0, r(\epsilon))} f(x).$$

This implies that

$$f(x_0) \leq \sup_{r>0} \inf_{x \in B(x_0, r)} f(x).$$

On the other hand, suppose that

$$f(x_0) \leq \sup_{r>0} \inf_{x \in B(x_0, r)} f(x).$$

Suppose to the contrary that f is not lower semicontinuous at x_0. This means that for every $t < f(x_0)$ and for every $r > 0$, there is a $\tilde{x} \in B(x_0, r)$ such that

$$f(\tilde{x}) < t.$$

For every $r > 0$, we can conclude that

$$\inf_{x \in B(x_0, r)} f(x) < t.$$

Hence, we have

$$t < f(x_0) \leq \sup_{r>0} \inf_{x \in B(x_0, r)} f(x) < t$$

which is a contradiction. $\qquad\square$

We can obtain a practical result that shows that lower semicontinuity of a functional on a metric space is in fact equivalent to sequential lower semicontinuity of the functional. This is not true on a general topological space.

Proposition 7.1.2. *Let (X, d) be a metric space. A functional $f : X \to \overline{\mathbb{R}}$ is sequentially lower semicontinuous at $x_0 \in X$ if*

$$f(x_0) \leq \liminf_{k} f(x_k)$$

for any sequence $\{x_k\}_{k \in \mathbb{N}}$ such that $x_k \to x_0$ in metric.

Theorem 7.1.2. *Let (X, d) be a metric space. A functional $f : X \to \overline{\mathbb{R}}$ is lower semi-continuous at $x_0 \in X$ if and only if f is sequentially lower semicontinuous at x_0.*

Thus, we see that on a metric space, we can characterize lower semicontinuous functionals by their behavior on convergent sequences. Now we show that a lower semicontinuous functional has a simple geometric interpretation. A lower semicontinuous function can be visualized in terms of its epigraph.

Theorem 7.1.3. *Let (X, d) be a metric space, and suppose that $f : X \to \overline{\mathbb{R}}$. The functional f is lower semicontinuous if and only if $\mathrm{epi}(f)$ is closed.*

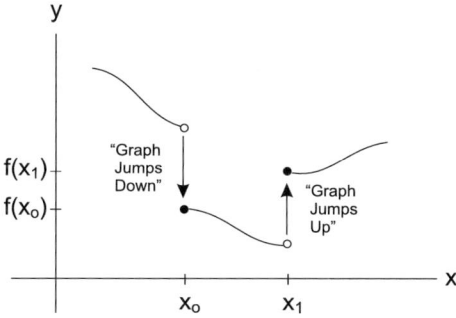

Figure 7.1: A function that is lower semicontinuous at x_0 but not lower semicontinuous at x_1.

Proof. Suppose that f is lower semicontinuous. Let $\{(x_k, \xi_k)\}_{k \in \mathbb{N}} \subseteq \mathrm{epi}(f)$, and that $(x_k, \xi_k) \to (x_0, \xi_0) \in X \times \mathbb{R}$. By the definition of the epigraph, we know that

$$f(x_k) \leq \xi_k \quad \forall k \in \mathbb{N}.$$

In particular, because $\{x_k\}$ converges to x_0, we can write

$$f(x_0) \leq \liminf_k f(x_k) \leq \liminf_k \xi_k = \lim_k \xi_k = \xi_0$$

since f is lower semicontinuous. This just means that $f(x_0) \leq \xi_0$, and that $(x_0, \xi_0) \in \mathrm{epi}(f)$. On the other hand, suppose that the $\mathrm{epi}(f)$ is closed. That is, if $\{(x_k, \xi_k)\}_{k \in \mathbb{N}} \subseteq \mathrm{epi}(f)$, then

$$(x_k, \xi_k) \to (x_0, \xi_0) \quad \text{in } X \times \mathbb{R}$$

implies that $(x_0, \xi_0) \in \mathrm{epi}(f)$. Consider any sequence such that

$$x_j \to x_0 \quad \in X.$$

Clearly, $\{(x_j, f(x_j))\}_{j \in \mathbb{N}}$ is contained in $\mathrm{epi}(f)$. Since the epigraph is closed, we can write

$$(x_j, f(x_j)) \to (x_0, \xi) \in \mathrm{epi}(f).$$

But now we get the required result

$$f(x_0) \leq \xi = \lim_j f(x_j) = \liminf_j f(x_j). \qquad \square$$

The graphical interpretation of this theorem is depicted in Figure 7.1. When the epigraph of f is closed, f makes no "jumps up" in its graph.

7.2 Lower Semicontinuous Functionals and Convexity

7.2.1 Banach Theorem for Lower Semicontinuous Functionals

We will show that convex lower semicontinuous functionals can always be minorized by affine functionals. This result is important for developing an intuition regarding the global behavior of lower semicontinuous functionals, as well as providing an important result for set-valued characterizations of differentiability.

Theorem 7.2.1 (Hahn-Banach Theorem for Lower Semicontinuous Functionals). *Let X be a normed vector space and suppose that the functional $f : X \to \overline{\mathbb{R}}$ is proper, convex and lower semicontinuous. Then f is bounded below by an affine functional.*

Recall that a functional is proper if it never takes the value $-\infty$ and there is at least one point at which the functional is finite. The proof of this theorem relies on several fundamental results that guarantee the separation of convex sets in normed vector spaces. The foundation of this proof is the *Hahn-Banach theorem*, one of the cornerstones of functional analysis. To prepare for the proof of this theorem, we will require the definition of hyperplane.

Definition 7.2.1. *A hyperplane $\mathcal{H}$ contained in a normed vector space X is the inverse image of a real number α under a given linear functional $x^* \in X^*$. In other words,*

$$\mathcal{H} = \{x \in X : \; < x^*, x >_{X^* \times X} = \alpha \} .$$

A hyperplane $\mathcal{H}$ is necessarily closed, because the inverse image of a closed set in the range of a continuous linear mapping is a closed set in the domain. A hyperplane has many uses, in particular, providing a notion of *separation* in a normed vector space.

Definition 7.2.2. *If $\mathcal{H}$ is a hyperplane defined on the normed vector space X, the open upper half-space is defined to be*

$$\mathcal{H}_+ = \{x \in X : \; < x^*, x >_{X^* \times X} > \alpha \} .$$

The open lower half-space is defined in an analogous manner:

Definition 7.2.3. *If $\mathcal{H}$ is a hyperplane defined on the normed vector space X, the open lower half-space is defined to be*

$$\mathcal{H}_- = \{x \in X : \; < x^*, x >_{X^* \times X} < \alpha \} .$$

With the definition of the upper and lower half-spaces, we can define the notion of separation of sets in a normed vector space.

Definition 7.2.4. *A hyperplane $\mathcal{H}$ strictly separates two sets $A, B \subseteq X$ if*

$$A \subseteq \mathcal{H}_+ \text{ and } B \subseteq \mathcal{H}_-.$$

In Chapter 2, we introduced the Hahn-Banach theorem and showed that it could be used to provide a rich collection of bounded linear functionals on normed vector spaces. These functionals were obtained via extension of functionals defined on subspaces. Sometimes, the version of the Hahn-Banach theorem discussed in Chapter 2 is referred to as the "extension version" of the Hahn-Banach theorem. It turns out that there are many forms of the Hahn-Banach theorem, and many of them are expressed in terms of separation of sets. The following theorem is one of the "separation versions" of the Hahn-Banach theorem.

Theorem 7.2.2. *Suppose that A and B are disjoint, convex subsets of a normed vector space X, where A is compact and B is closed. Then there is a hyperplane that strictly separates A and B.*

Now we can return to the proof of Theorem 7.2.1.

Proof. Choose some value $t \in \mathbb{R}$ such that

$$t < f(x_0)$$

for some $x_0 \in X$. Since f is proper, we know that some such x_0 exists. The set $\{x_0, t\}$ is a compact, convex subset of $X \times \mathbb{R}$, and it is disjoint from the epigraph of f. Since f is lower semicontinuous and convex, its epigraph epi(f) is closed and convex. By the Hahn-Banach Theorem 7.2.2, there is a functional $g \in (X \times \mathbb{R})^*$ such that

$$\langle g, (x_0, t) \rangle_{(X \times \mathbb{R})^* \times (X \times \mathbb{R})} < \alpha < \langle g, (x, s) \rangle_{(X \times \mathbb{R})^* \times (X \times \mathbb{R})}$$

for all $(x, s) \in$ epi(f). Because $(X \times \mathbb{R})^* = X^* \times \mathbb{R}$, each $g \in (X \times \mathbb{R})^*$ can be written as

$$g = (x^*, \tau)$$

where $x^* \in X^*$ and $\tau \in \mathbb{R}$. In this case, the statement of separation can be written as

$$\langle x^*, x_0 \rangle_{X^* \times X} + \tau t < \alpha < \langle x^*, x \rangle_{X^* \times X} + \tau s \quad \forall (x, s) \in \text{epi}(f).$$

Since $(x_0, f(x_0)) \in$ epi(f), the last inequality is reduced to

$$\langle x^*, x_0 \rangle_{X^* \times X} + \tau t < \alpha < \langle x^*, x_0 \rangle_{X^* \times X} + \tau f(x_0)$$

which implies that

$$0 < \tau \underbrace{(f(x_0) - t)}_{>0}.$$

Consequently, we have $\tau > 0$. Finally,

$$\frac{\alpha}{\tau} - \frac{1}{\tau}\langle x^*, x\rangle_{X^* \times X} < s \quad \forall\, (x, s) \in \mathrm{epi}(f).$$

But since $(x, f(x)) \in \mathrm{epi}(f)$, the last inequality means that the convex lower semicontinuous functional f is minorized by an affine functional

$$\frac{\alpha}{\tau} - \frac{1}{\tau}\langle x^*, x\rangle_{X^* \times X} < f(x). \qquad \square$$

7.2.2 Gateaux Differentiability

In our consideration of the relationship of convexity, differentiability and lower semicontinuity, we show that Gateaux differentiability is also useful in proving that a functional is weakly sequentially lower semicontinuous.

Theorem 7.2.3. *Suppose X is a separable, reflexive Banach space and that $f : M \subseteq X \to \overline{\mathbb{R}}$ is Gateaux differentiable over the closed, convex set M. Then if f is convex, f is weakly sequentially lower semicontinuous.*

Proof. Consider $x, y \in M$. By the convexity of f, we can write

$$\begin{aligned}
\alpha f(x) + (1 - \alpha)f(y) &\geq f(\alpha x + (1 - \alpha)y) \\
f(x) - f(y) &\geq \frac{1}{\alpha}\{f(y + \alpha(x - y)) - f(y)\} \quad \forall\, x, y \in M \\
f(x) - f(y) &\geq \langle Df(y), x - y\rangle_{X^* \times X} \quad \forall\, x, y \in M.
\end{aligned}$$

Now choose any sequence such that $x_n \to x$ weakly in X. We can write

$$\begin{aligned}
f(x_n) - f(y) &\geq \langle Df(y), x_n - y\rangle_{X^* \times X} \quad \forall\, y \in M \ \forall\, n \in \mathbb{N} \\
\liminf_n (f(x_n) - f(y)) &\geq \liminf_n \langle Df(y), x_n - y\rangle_{X^* \times X} \quad \forall\, y \in M \\
&= \lim_n \langle Df(y), x_n - y\rangle_{X^* \times X} \quad \forall\, y \in M.
\end{aligned}$$

If we choose $y = x$ in the above expression, we obtain

$$\liminf_n \big(f(x_n) - f(x)\big) \geq \lim_n \langle Df(x), x_n - x\rangle_{X^* \times X} = 0.$$

But this means that

$$f(x) \leq \liminf_n f(x_n). \qquad \square$$

7.2.3 Lower Semicontinuity in Weak Topologies

If we can show that a functional is convex, we can choose the weak or strong topology to study lower semicontinuity. In the following theorem, we are able to state conditions under which lower semicontinuity and *lower semicontinuity in the weak topology* are established. Interestingly enough, these two notions of semicontinuity can both be shown to hold if the functional f is convex.

Definition 7.2.5. *Let (X, τ) be a topological space. A functional $f : X \to \overline{\mathbb{R}}$ is τ-lower semicontinuous on X if and only if* $\mathrm{epi}(f)$ *is τ-closed in the vector space $X \times \mathbb{R}$.*

We renew a well-known relationship between weak and strong closures at convex sets.

Theorem 7.2.4 (Mazur's Theorem). *A convex subset of a Banach space is strongly closed if and only if it is weakly closed.*

Theorem 7.2.5. *Suppose that X is a Banach space and $f : X \to \overline{\mathbb{R}}$ is convex. Then f is lower semicontinuous in the strong topology on X if and only if f is lower semicontinuous in the weak topology on X.*

Proof. Recall, that if $\mathrm{epi}(f)$ is lower semicontinuous in the strong topology on X, $\mathrm{epi}(f)$ is strongly closed. Since f is convex, $\mathrm{epi}(f)$ is also convex. But a convex set is strongly closed in a Banach space if and only if it is weakly closed. Hence, $\mathrm{epi}(f)$ is closed in the weak topology in X, and therefore lower semicontinuous in the weak topology on X. $\qquad\square$

Now in many applications, it is not the case that the functional of interest is defined on the entire space X. Generally, the domain of definition of f will be subject to constraints. If the constraint set is closed and convex, however, the conclusions of the preceding theorem still hold true.

Theorem 7.2.6. *Suppose that X is a Banach space, $f : M \subseteq X \to \overline{\mathbb{R}}$ is convex, and M is closed and convex. Then f is strongly lower semicontinuous if and only if f is weakly lower semicontinuous.*

Proof. Recall that a function is τ-lower semicontinuous if and only if all of its lower sections $M_r = \{x \in M : f(x) \leq r\}$ are τ-closed for all $r \in \mathbb{R}$. Consequently, if f is strongly lower semicontinuous, then M_r is strongly closed in X. Since f is convex, M_r is convex for each $r \in \mathbb{R}$. Indeed, suppose that $x, y \in M_r$. By the convexity of f and $f(x)$ and $f(y)$ boundedness above by r when $x, y \in M_r$, we have

$$f(\alpha x + (1 - \alpha)y) \leq \alpha f(x) + (1 - \alpha)f(y) \leq \alpha r + (1 - \alpha)r = r.$$

Note that $\alpha x + (1 - \alpha)y \in M$ since M is convex. Hence, M_r is convex and strongly closed, and therefore, by Mazur's Theorem 7.2.4, M_r is weakly closed. But this means that f is weakly lower semicontinuous. $\qquad\square$

Now, to summarize the development thus far, we have shown that:

- concise existence results for minimization problems are available for lower semicontinuous functionals,
- lower semicontinuous functionals have a simple geometric interpretation in terms of the epigraph, and
- convexity widens the class of topologies for which lower semicontinuity can be deduced.

7.3   The Generalized Weierstrass Theorem

The greatest difficulty encountered in applying the conditions of sufficiency stated in Theorem 5.3.2 is that nth-order Fréchet differentiability is a restrictive assumption indeed. Moreover, the task of verifying that an nth-order Fréchet derivative is bounded away from zero, or coercive, is a difficult task in itself. In this section, and the next, we discuss how the assumptions on the functional f can be weakened, so that we still can derive conditions that guarantee a solution of the original minimization problem.

Now we can state the generalized Weierstrass Theorem, which forms the basis of nearly all of the sufficiency conditions we employ in the remainder of this volume. The generalized Weierstrass Theorem weakens the requirement of continuity, to that of *lower semicontinuity*.

Theorem 7.3.1 (Generalized Weierstrass Theorem). *Let (X, τ) be a compact topological space, and let $f : X \to \overline{\mathbb{R}}$ be a proper, lower semicontinuous functional. Then there exists $x_0 \in X$ such that*

$$f(x_0) = \inf_{x \in X} f(x).$$

The proof of this theorem will closely follow the proof of the Weierstrass Theorem.

Proof. Suppose that $\alpha = \inf_{x \in X} f(x)$. Consider the sequence of half-closed sets in the range of f

$$Q_k = \left(-\infty, \alpha + \frac{1}{k}\right] \quad \forall\, k \in \mathbb{N}.$$

From this sequence of sets, we can construct a sequence of closed sets in the domain X

$$C_k = \{x \in X : f(x) \in Q_k\} \quad \forall\, k \in \mathbb{N}.$$

By construction, this sequence of sets is nested

$$C_{k+1} \subseteq C_k \quad \forall\, k \in \mathbb{N}$$

and each C_k is compact, being a closed subset of a compact set. Following identical reasoning as in the proof of Theorem 5.1.1, we conclude that

$$\exists\, x_0 \in \bigcap_{k=1}^{\infty} C_k$$

and, therefore,

$$f(x_0) = \alpha = \inf_{x \in X} f(x). \qquad \square$$

Now, admittedly, the preceding discussions of continuity and lower semicontinuity are expressed in somewhat abstract form. Checking to see if a functional to be minimized "maps half-closed sets into half-closed sets" is not a task that would be embraced enthusiastically by someone interested in a particular applications. In the remainder of this section, we investigate lower semicontinuity in more detail, and study its relation to convexity and differentiability.

7.3.1 Compactness in Weak Topologies

Up until this point we have emphasized that there are two key ingredients in Theorem 7.3.1: the lower semicontinuity of the functional f and the compactness of the underlying space. We have discussed the role of convexity in establishing lower semicontinuity in either the strong or weak topologies. Thus, convexity provides a practical means of establishing lower semicontinuity in many applications. In this section, we discuss a framework for addressing the second primary hypothesis of Theorem 7.3.1: compactness. As we recall from Chapter 2, several results are available that describe conditions that ensure weak compactness in various scenarios. Alaouglu's theorem is perhaps the most general and guarantees the weak* compactness of the closed unit ball of the topological dual space of a Banach space.

Theorem 7.3.2. *The closed unit ball of the dual space of a Banach space is weakly* compact.*

From this quite general result, it has been shown in Section 1.5 that alternative, more applicable theorems may be derived. The following theorem is used frequently throughout various applications in mechanics and optimal control.

Theorem 7.3.3. *A bounded sequence in a reflexive Banach space contains a weakly convergent subsequence.*

This theorem follows from the previous one essentially because the weak and weak* topologies on the topological dual space of a Banach space coincide. Carefully note that this theorem says that a bounded sequence of a reflexive Banach space is *almost* weakly sequentially compact. The weakly convergent subsequence guaranteed by the theorem need not converge to an element of the bounded subset.

These two theorems allow us to construct what will be the foundation of the most frequently used technique for employing Theorem 7.3.1 in practical applications. As a general philosophy, we

   (i) seek to establish lower semicontinuity in the weak topology on X, and
   (ii) seek to establish compactness in the weak topology on X.

From the foregoing comments, boundedness is nearly sufficient for weak sequential compactness of a subsequence in a reflexive Banach space. Some care must be taken as we proceed. In deriving the conditions that follow, it is important to note that every weakly lower semicontinuous functional is automatically weakly sequentially lower semicontinuous. This fact follows, for example, from Property

(IV) in Theorem 7.1.1. It is not true in a general topological space, however, that a sequentially lower semicontinuous functional is lower semicontinuous. The equivalence of sequential lower semicontinuity and lower semicontinuity holds if the topological space is first countable (has a countable base for the neighborhood system at each point). See [19, Proposition 1.3, p. 9]. Similarly, every set that is closed in the weak topology is weakly sequentially closed. This fact follows from Theorem 1.2.8. However, it is not the case that every sequentially closed set is a closed set in a general topological space. We can now write a version of the generalized Weierstrass Theorem that is tailored to our approach.

Theorem 7.3.4. *Let X be a reflexive Banach space, and let $f : M \subseteq X \to \overline{\mathbb{R}}$ be weakly sequentially lower semicontinuous over the bounded and weakly sequentially closed subset M. Then, $\exists x_0 \in M$ such that*

$$f(x_0) = \inf_{x \in M} f(x).$$

Proof. We first show that f is bounded below. Suppose to the contrary that there is a sequence $\{x_k\}_{k \in \mathbb{N}} \subseteq M$ such that

$$f(x_k) < -k.$$

Since M is bounded, there is a subsequence $\{x_{k_j}\}_{j \in \mathbb{N}}$ such that

$$x_{k_j} \to \overline{x}$$

weakly in X. Because M is weakly sequentially closed, $\overline{x} \in M$. By the weak sequential lower semicontinuity of f

$$f(\overline{x}) \leq \liminf_j f(x_{k_j}) = -\infty.$$

But this is a contradiction and f is bounded below. Let

$$\alpha = \inf_{x \in M} f(x)$$

and suppose $\{x_k\}_{k \in \mathbb{N}}$ is a sequence contained in M such that

$$\alpha = \lim_k f(x_k).$$

Since M is bounded and weakly sequentially closed, there is a subsequence $\{x_{k_j}\}_{j \in \mathbb{N}} \subseteq M$ such that

$$x_{k_j} \to x_0 \in M$$

weakly in X. By the weak sequential lower semicontinuity of f, we can write

$$\alpha \leq f(x_0) \leq \liminf_j f(x_{k_j}) = \lim_j f(x_{k_j}) = \alpha$$

and the theorem is proved.   □

To derive more applicable results, we would like to bring to bear all of the tools we derived in Sections 6.2 and 7.2.2 that employ the relationship among *convexity, differentiability* and *lower semicontinuity.*

7.3.2 Bounded Constraint Sets

One direct result of Theorem 7.3.4 is the following theorem, that is applicable when the constraint set is bounded.

Theorem 7.3.5. *Let X be a reflexive Banach space and suppose that $f : M \subseteq X \to \overline{\mathbb{R}}$ is weakly sequentially lower semicontinuous over the bounded, convex, closed subset M. Then, $\exists\, x_0 \in X$ such that*

$$f(x_0) = \inf_{x \in M} f(x).$$

Proof. Since M is closed and convex, it is closed in the weak topology on X. It is therefore weakly sequentially closed. The conclusion now follows from Theorem 7.3.4. $\qquad\square$

If it happens that we can show that f is Gateaux differentiable, the following theorem provides several useful alternatives for directly showing that the minimization problem has a solution.

Theorem 7.3.6. *Let X be a reflexive Banach space and suppose that $f : M \subseteq X \to \overline{\mathbb{R}}$ is Gateaux differentiable on the closed, convex and bounded subset M. If any of the following three conditions holds true,*

(i) *f is convex over M,*
(ii) *Df is monotone over M,*
(iii) *$D^2 f$ is positive over M,*

all three conditions hold, and $\exists\, x_0 \in X$ such that

$$f(x_0) = \inf_{x \in M} f(x).$$

Proof. By Theorem 6.2.1 and Theorem 7.2.3, f is weakly sequentially lower semicontinuous under any of the hypothesis (i) through (iii) above. The proof now follows from Theorem 7.3.5. $\qquad\square$

7.3.3 Unbounded Constraint Sets

In the preceding section, we saw that by considering *bounded* constraint sets, utilizing convexity and by employing the weak topology on X, the minimization problem of finding $x_0 \in M$ such that

$$f(x_0) = \inf_{x \in M} f(x)$$

has a solution. For a surprisingly large class of problems, we can "bootstrap" this argument to treat some *unbounded constraint* sets. The crucial additional property we impose when the constraint set is not necessarily bounded is that the functional f is *coercive*.

Definition 7.3.1. *Let X be a normed vector space and let*

$$f : M \subseteq X \to \mathbb{R}.$$

The functional f is coercive over M if

$$\|x\| \to \infty \quad \Longrightarrow \quad |f(x)| \to \infty$$

for all $x \in M$.

How we will exactly use this condition becomes clearer when we note that if f is coercive and bounded over a set M, then the set M itself is bounded. For example, if f is coercive, then for two constants c_1 and c_2, we have

$$|f(x)| \le c_1 \;\; \forall x \in M \quad \Longrightarrow \quad \|x\| \le c_2 \;\; \forall x \in M.$$

We derive the following supporting proposition.

Proposition 7.3.1. *Let X be a normed vector space, M be a closed subset of X, and let $f : M \subseteq X \to \mathbb{R}$ be coercive. Then $\exists z \in M$ and $\exists R > 0$ such that*

$$\inf_{x \in M} f(x) = \inf \left\{ f(x) : x \in M \cap \overline{B_R(z)} \right\}.$$

Proof. Pick any $z \in M$. We claim that there is an $R > 0$ such that the condition

$$\|x - z\| > R \quad \text{implies} \quad |f(x)| \ge |f(z)| \tag{7.1}$$

for all $x \in M$. Suppose to the contrary that this is not the case. Then for every $R > 0$ there is an $x_R \in M$ such that

$$\|x_R - z\| > R \quad \text{and} \quad |f(x_R)| < |f(z)|.$$

But as $R \to \infty$, we have

$$\|x_R\| + \|z\| \ge \|x_R - z\| > R$$

and therefore

$$\|x_R\| \to \infty.$$

On the other hand,

$$|f(x_R)| < |f(z)|$$

is a contradiction of the coercivity of f. The claim in Equation (7.1) follows. The proposition now follows since

$$\inf\{f(x) : x \in M\} = \inf \left\{ f(x) : x \in \left(M \cap \overline{B_R(z)} \right) \cup \left(M \backslash \overline{B_R(z)} \right) \right\}.$$

But since

$$f(z) \le \inf\{f(x) : x \in M \backslash \overline{B_R(z)}\}$$

we conclude that

$$\inf\{f(x) : x \in M\} = \inf\{f(x) : x \in M \cap \overline{B_R(z)}\}. \qquad \square$$

Theorem 7.3.7. *Let X be a reflexive Banach space, and let $f : M \subseteq X \to \mathbb{R}$ be coercive and weakly sequentially lower semicontinuous over the closed and convex set M. Then $\exists\, x_0 \in X$ such that*

$$f(x_0) = \inf_{x \in M} f(x).$$

Proof. Based on Proposition 7.3.1, the proof of the theorem is straightforward. By Proposition 7.3.1, we choose $R > 0$ and $z \in M$ such that

$$\inf_{x \in M} f(x) = \inf_{x \in M \cap \overline{B_R(z)}} f(x).$$

The ball $\overline{B_R(z)}$ is closed, convex and bounded. Its intersection with M is likewise closed, convex and bounded in a reflexive Banach space. Thus, we can apply Theorem 7.3.5. $\square$

We can view this theorem as a prototypical minimization result for unbounded constraint sets. We can ensure that the hypotheses of this theorem are satisfied by bringing to bear all of the tests for convexity, monotonicity of the first Gateaux differential, and positivity of the second Gateaux differential discussed in Sections 6.2 and 7.2.2.

Theorem 7.3.8. *Let X be a reflexive Banach space and let $f : M \subseteq X \to \overline{\mathbb{R}}$ be Gateaux differentiable and coercive on the nonempty, closed, convex set M. If one of the following conditions hold,*

(i) *f is convex,*
(ii) *Df is monotone, or*
(iii) *$D^2 f$ is positive,*

then all of the conditions (i) through (iii) hold, and $\exists\, x_0 \in X$ such that

$$f(x_0) = \inf_{x \in M} f(x).$$

Proof. The proof follows from noting that if f is convex and Gateaux differentiable, then it is weakly sequentially lower semicontinuous by Theorem 7.2.3. In addition, Theorem 6.2.1 notes that, under the hypotheses of this theorem, the convexity of f in condition (i) is equivalent to the monotonicity of the first differential in (ii) or the positivity of the second differential in (iii). The proof now follows from Theorem 7.3.7. $\square$

7.3.4 Constraint Sets on Ordered Vector Spaces

In this section we consider problems that have a structure quite similar to that considered in Section 6.4. Suppose $f : \mathcal{U} \subseteq X \to \mathbb{R}$ and $g : \mathcal{U} \subseteq X \to Y$, where X and Y are normed vector spaces and $\mathcal{U}$ is a convex subset of X. We study the optimization problem in which we seek $x_0 \in \mathcal{U} \subseteq X$ such that

$$f(x_0) = \inf\{f(x) : g(x) \leq 0,\ x \in \mathcal{U}\} \tag{7.2}$$

and
$$g(x_0) \leq 0.$$

The normed vector space Y contains a cone P, that induces an ordering of Y. Hence, expressions like
$$g(x) \leq 0$$
are understood in the sense discussed in Section 6.4.1. That is, if y_1, $y_2 \in Y$

$$y_2 \geq y_1 \quad \Longleftrightarrow \quad y_2 - y_1 \in P.$$

Section 6.5 has discussed several techniques by which the existence of a solution to this optimization problem can be established. The following theorem is but a slight generalization of those results to accommodate convex constraints.

Theorem 7.3.9. *Suppose that*

- *Y is a normed vector space with a closed convex cone P,*
- *$f : X \rightarrow \mathbb{R}$ is weakly sequentially lower semicontinuous,*
- *g is weakly sequentially continuous,*
- *the optimization problem (7.2) is feasible, and*
- *$\mathcal{U}$ is weakly sequentially compact.*

Then $\exists\, x_0$ such that

$$f(x_0) = \inf\{f(x) : g(x) \leq 0,\ x \in \mathcal{U} \subseteq X\}$$

and

$$g(x_0) \leq 0.$$

Proof. The proof of this theorem follows closely the methodology employed in the derivation of Theorem 7.3.7. The only significant difference is the form of the constraint in the current theorem. By the definition of the infimum,

$$\exists\{x_k\}_{k \in \mathbb{N}} \subseteq \mathcal{U}$$

such that

$$f(x_k) \rightarrow f(x_0)$$

and

$$g(x_k) \leq 0 \quad \forall\, k \in \mathbb{N}.$$

Since $\mathcal{U}$ is weakly sequentially compact, there is a subsequence

$$\{x_{k_j}\}_{j \in \mathbb{N}} \subseteq \{x_k\}_{k \in \mathbb{N}} \subseteq \mathcal{U}$$

such that

$$x_{k_j} \rightarrow x_0 \quad \text{weakly in} \quad X.$$

Since $\mathcal{U}$ is weakly sequentially compact, it is weakly sequentially closed and $x_0 \in \mathcal{U}$. By the weak sequential lower semicontinuity of f, we conclude

$$f(x_0) \leq \liminf_k f(x_k) = \lim_k f(x_k)$$

and, consequently, obtain

$$f(x_0) = \inf\{f(x) : g(x) \leq 0, \ x \in \mathcal{U} \subseteq X\}.$$

It only remains to show that $g(x_0) \leq 0$. However, since g is weakly sequentially continuous, we have

$$g(x_{k_j}) \to g(x_0).$$

Because $g(x_{k_j}) \in P$, and P is a weakly sequentially closed set, we conclude that $g(x_0) \in P$. □

   The reader should note that numerous variants of this theorem are possible. We can modify the hypotheses to guarantee weak sequential continuity and weak sequential compactness. For example, the conclusions of the theorem hold if f is Gateaux differentiable and convex.

References

[1] R.A. Adams. *Sobolev Spaces*. Academic Press, New York, 1975.

[2] J. Bergh and J. Lofstrom. *Interpolation Space: An Introduction*. Springer, Berlin, 1976.

[3] A. Brown and C. Pearcy. *Introduction to Operator Theory: Elements of Functional Analysis*. Springer, New York, 1977.

[4] P.G. Ciarlet. *The Finite Element Methods for Elliptic Problems*. North Holland Publishing Company, New York, 1978.

[5] J.L. Doob. *Measure Theory*. Springer, New York, 1994.

[6] R.E. Edwards. *Functional Analysis: Theory and Applications*. Dover Publications, Inc., New York, 1995.

[7] I. Ekeland and R. Temam. *Convex Analysis and Variational Problems*. North-Holland Publishing, Amsterdam, 1976.

[8] P.R. Halmos. *Measure Theory*. Springer, New York, 1997.

[9] E. Hewitt and K. Stromberg. *Real and Abstract Analysis*. Springer, Berlin, 1965.

[10] F. Hirsch and G. Lacombe. *Elements of Functional Analysis*. Springer, New York, 1999.

[11] A. Ioffe and V. Tichomirov. *Theory of Extremal Problems*. North Holland, Amsterdam, 1979.

[12] W. Kaplan. *Advanced Calculus*. Addison-Wesley, 2002.

[13] N. Kikuchi and J. Tinsley Oden. *Contact Problems in Elasticity*. Society for Industrial and Applied Mathematics Publications, Philadephia, 1982.

[14] A.N. Kolmogorov and S.V. Fomin. *Introductory Real Analysis*. Dover Publications, Inc., 1970.

[15] E. Kreyszig. *Introductory Functional Analysis with Applications*. John Wiley & Sons, Inc., 1978.

[16] A. Kufner, O. John, and S. Fucik. *Function Spaces*. Noordhoff International Publishing, Prague, 1977.

[17] D.G. Luenberger. *Optimization by Vector Space Methods*. John Wiley & Sons, Inc., 1969.

[18] J.E. Marsden and T.J.R. Hughes. *Mathematical Foundations of Elasticity*. Dover Publications, Inc., 1983.

[19] G. Dal Maso. *An Introduction to Γ-convergence*. Birkhäuser, 1993.

[20] R.E. Megginson, A. Sheldon, and F.W. Gehring. *An Introduction to Banach Space Theory*. Springer, New York, 1998.

[21] J.T. Oden. *Qualitative Methods in Nonlinear Mechanics*. Prentice-Hall, Inc., 1986.

[22] P.J. Olver. *Applications of Lie Groups to Differential Equations*. Springer, New York, 1993.

[23] P. Pedregal. *Parametrized Measures and Variational Principles*. Birkhäuser, Boston, 1997.

[24] M. Renardy and R.C. Rogers. *An Introduction to Partial Differential Equations*. Springer, New York, 1992.

[25] R.T. Rockafellar. *Convex Analysis*. Princeton University Press, Princeton, N.J, 1970.

[26] T. Roubicek. *Relaxation in Optimization Theory and Variational Calculus*. Walter de Gruyter Publisher, New York, 1997.

[27] H.L. Royden. *Real Analysis*. MacMillan Publishing Co., Inc., 1968.

[28] W. Rudin. *Functional Analysis*. McGraw-Hill, Inc., 1973.

[29] H.J. Schmeisser and H. Triebel. *Topics in Fourier Analysis and Function Spaces*. Akadem. Verlagsgesellschaft Geest & Portig, Leipzig, 1987.

[30] A.E. Taylor and D.C. Lay. *Introduction to Functional Analysis*. John Wiley & Sons, Inc., New York-Chichester-Brisbane, 1980.

[31] H. Triebel. *Theory of Function Spaces*. Birkhäuser, Leipzig, 1983.

[32] J.L. Troutman. *Variational Calculus and Optimal Control*. Springer, New York, 1995.

[33] J. Warga. *Optimal Control of Differential Equations and Functional Equations*. Academic Press, New York, 1972.

[34] A. Wilansky. *Topology for Analysis*. Robert E. Krieger Publishing Company, Inc., 1983.

[35] J. Wloka. *Partial Differential Equations*. Cambridge University Press, 1987.

[36] W.M. Wonham. *Linear Multivariable Control: A Geometric Approach*. Springer, New York, 1985.

[37] E. Zeidler. *Nonlinear Functional Analysis and Its Applications: Volume (II/A) Linear Monotone Operators*. Springer, New York, 1990.

Index

Systems and Control: Foundations and Applications

A series of monographs and advanced graduate texts

Edited by
Başar, T., Washington University, St. Louis, USA

Systems and Control is designed for the publication of research level monographs and advanced graduate textbooks in all areas of systems and control theory and its applications to a wide variety of scientific disciplines.

Your Specialized Publisher in Mathematics
Birkhäuser

For orders originating from all over the world except USA/Canada/Latin America:

Birkhäuser Verlag AG
c/o Springer GmbH & Co
Haberstrasse 7
D-69126 Heidelberg
Fax: +49 / 6221 / 345 4 229
e-mail: birkhauser@springer.de
http://www.birkhauser.ch

For orders originating in the USA/Canada/Latin America:

Birkhäuser
333 Meadowland Parkway
USA-Secaucus
NJ 07094-2491
Fax: +1 201 348 4505
e-mail: orders@birkhauser.com

Abou-Kandil, H. / Freiling, G. / Ionescu, V. / Jank, G., Matrix Riccati Equations in Control and Systems Theory (2003)
ISBN 3-7643-0085-X

Allen, J.C., H^∞ Engineering and Amplifier Optimization (2005)
ISBN 0-8176-3780-X

Asachenkov, A. / Marchuk, G. / Mohler, R. / Zuev, S., Disease Dynamics (1994)
ISBN 0-8176-3692-7

Aubin, J.P., Mutational and Morphological Analysis. Tools for Shape Evolution and Morphogenesis (1998)
ISBN 0-8176-3935-7

Aubin, J.P., Viability Theory (1991)
ISBN 0-8176-3571-8

Bardi, M. / Capuzzo Dolcetta, I., Optimal Control and Viscosity Solutions of Hamilton-Jacobi-Bellman Equations (1998)
ISBN 0-8176-3640-4

Bashirov, A.E., Partially Observable Linear Systems Under Dependent Noises (2003)
ISBN 3-7643-6999-X

Bensoussan, A. / Da Prato, G. / Delfour, M.C. / Mitter, S.K., Representation and Control of Infinite Dimensional Systems. Volume 1 (1992)
ISBN 0-8176-3641-2
Volume 2 (1993)
ISBN 0-8176-3642-0

Byrnes, C.I. / DelliPriscoli, F. / Isidori, A., Output Regulation of Uncertain Nonlinear Systems (1997)
ISBN 0-8176-3997-7

Chen, H.F. / Guo, L., Identification and Stochastic Adaptive Control (1992)
ISBN 0-8176-3597-1

Christofides, P.D., Nonlinear and Robust Control of PDE Systems. Methods and Applications to Transport-Reaction Processes (2001)
ISBN 0-8176-4156-4

Colonius, F. / Kliemann, W., The Dynamics of Control. With an Appendix by Lars Grüne (2000)
ISBN 0-8176-3683-8

Colonius, F. / Wirth, F. / Helmke, U. / Prätzel-Wolters, D., Advances in Mathematical Systems Theory. A Volume in Honor of Diederich Hinrichsen (2000). ISBN 0-8176-4162-9

Systems and Control: Foundations and Applications

Your Specialized Publisher in Mathematics

Birkhäuser

Davis, J.H., Foundations of Deterministic and Stochastic Control (2002)
ISBN 0-8176-4257-9

Dragan, V. / Halanay, A., Stabilization of Linear Systems (1999)
ISBN 0-8176-3970-5

Dullerud, G., Control of Uncertain Sampled-Data Systems (1996)
ISBN 0-8176-3851-2

Feuer, A. / Goodwin, G.C., Sampling in Digital Signal Processing and Control (1996)
ISBN 0-8176-3934-9

van Keulen, B., H^∞ Control for Distributed Parameter Systems: A State-Space Approach (1993)
ISBN 0-8176-3709-5

Kimura, H., Chain-Scattering Approach to H^∞-Control (1997)
ISBN 0-8176-3787-7

Krasovskii, A.N. / Krasovskii, N.N., Control and Lack of Information (1995)
ISBN 0-8176-3698-6

Kuijper, M., First-order Representations of Linear Systems (1994)
ISBN 0-8176-3754-0

Kurdila, A. / Zabarankin, M., Convex Functional Analysis and Applications (2005)
ISBN 3-7643-2198-9

Kurzhanski, A. / Vályi, I., Ellipsoidal Calculus for Estimation and Control (1996)
ISBN 0-8176-3699-4

Lagnese, J.E. / Leugering, G. / Schmidt, E.J., Modeling, Analysis and Control of Dynamic Elastic Multi-Link Structures (1994)
ISBN 0-8176-3705-2

Li, X. / Yong , J., Optimal Control Theory for Infinite Dimensional Systems (1995)
ISBN 0-8176-3722-2

Liberzon, D., Switching in Systems and Control (2003)
ISBN 0-8176-4297-8

Mareels, I. / Polderman, J.W., Adaptive Systems: An Introduction (1996)
ISBN 0-8176-3877-6

McEneaney, W. / Yin, G. / Zhang, Q., Stochastic Analysis, Control, Optimization and Applications. A Volume in Honor of W.H. Fleming (1998)
ISBN 0-8176-4078-9

Peters, M.A. / Iglesias P.A., Minimum Entropy Control for Time-Varying Systems (1997)
ISBN 0-8176-3972-1

Sethi, S.P. / Zhang, Q., Hierarchical Decision Making in Stochastic Manufacturing Systems (1994)
ISBN 0-8176-3735-4

Srikant, R., The Mathematics of Internet Congestion Control (2003)
ISBN 0-8176-3227-1

Subrahmanyam, M.B., Finite Horizon H^∞ and Related Control Problems (1995)
ISBN 0-8176-3811-3

Tay, T.-T. / Mareels, I. / Moore, J.B., High Performance Control (1997)
ISBN 0-8176-4004-5

Vinter, R., Optimal Control (2000)
ISBN 0-8176-4075-4

Zabczyk, J., Mathematical Control Theory. An Introduction (1995) (2. printing)
ISBN 0-8176-3645-5